SpringerBriefs in Computer Science

Series editors

Stan Zdonik
Peng Ning
Shashi Shekhar
Jonathan Katz
Xindong Wu
Lakhmi C. Jain
David Padua
Xuemin Shen
Borko Furht
V. S. Subrahmanian
Martial Hebert
Katsushi Ikeuchi
Bruno Siciliano

For further volumes:
http://www.springer.com/series/10028

Wyatt Travis Clark

Information-Theoretic Evaluation for Computational Biomedical Ontologies

 Springer

Wyatt Travis Clark
Department of Molecular Biophysics
  and Biochemistry
Yale University
New Haven, CT
USA

ISSN 2191-5768 ISSN 2191-5776 (electronic)
ISBN 978-3-319-04137-7 ISBN 978-3-319-04138-4 (eBook)
DOI 10.1007/978-3-319-04138-4
Springer Cham Heidelberg New York Dordrecht London

Library of Congress Control Number: 2013957360

© The Author(s) 2014
This work is subject to copyright. All rights are reserved by the Publisher, whether the whole or part of the material is concerned, specifically the rights of translation, reprinting, reuse of illustrations, recitation, broadcasting, reproduction on microfilms or in any other physical way, and transmission or information storage and retrieval, electronic adaptation, computer software, or by similar or dissimilar methodology now known or hereafter developed. Exempted from this legal reservation are brief excerpts in connection with reviews or scholarly analysis or material supplied specifically for the purpose of being entered and executed on a computer system, for exclusive use by the purchaser of the work. Duplication of this publication or parts thereof is permitted only under the provisions of the Copyright Law of the Publisher's location, in its current version, and permission for use must always be obtained from Springer. Permissions for use may be obtained through RightsLink at the Copyright Clearance Center. Violations are liable to prosecution under the respective Copyright Law.
The use of general descriptive names, registered names, trademarks, service marks, etc. in this publication does not imply, even in the absence of a specific statement, that such names are exempt from the relevant protective laws and regulations and therefore free for general use.
While the advice and information in this book are believed to be true and accurate at the date of publication, neither the authors nor the editors nor the publisher can accept any legal responsibility for any errors or omissions that may be made. The publisher makes no warranty, express or implied, with respect to the material contained herein.

Printed on acid-free paper

Springer is part of Springer Science+Business Media (www.springer.com)

Preface

The development of effective methods for the prediction of ontological annotations is an important goal in computational biology, with protein function prediction and disease gene prioritization gaining wide recognition. While various algorithms have been proposed for these tasks, evaluating their performance is difficult due to problems caused both by the structure of biomedical ontologies and biased or incomplete experimental annotations of genes and gene products. In this work, we propose an information-theoretic framework to evaluate the performance of computational protein function prediction. We use a Bayesian network, structured according to the underlying ontology, to model the prior probability of a protein's function. We then define two concepts, misinformation and remaining uncertainty, that can be seen as information-theoretic analogs of precision and recall. Finally, we propose a single statistic, referred to as semantic distance, that can be used to rank classification models. We evaluate our approach by analyzing the performance of three protein function predictors of Gene Ontology terms and provide evidence that we address several weaknesses of currently used metrics. We believe this framework provides valuable and useful insights into the performance of protein function prediction tools.

Contents

1 **Introduction** 1
   1.1 Background 4
   1.2 Protein Function Prediction Scenarios 6
   1.3 State of the Art Methods 7
   References 7

2 **Methods** 13
   2.1 Calculating the Joint Probability of a Graph 13
       2.1.1 Calculating the Information Content of a Graph 16
       2.1.2 Comparing Two Annotation Graphs 17
       2.1.3 Measuring the Quality of Function Prediction 18
       2.1.4 Weighted Metrics 20
       2.1.5 Semantic Distance 20
       2.1.6 Precision and Recall 21
       2.1.7 Supplementary Evaluation Metrics 22
       2.1.8 Additional Topological Metrics 25
   2.2 Confusion Matrix Interpretation of *ru* and *mi* 25
   2.3 Annotation Models 26
       2.3.1 The Naïve Model 26
       2.3.2 The BLAST Model 27
       2.3.3 The GOtcha Model 27
   References 27

3 **Experiments and Results** 29
   3.1 Average Information Content of a Protein 29
   3.2 Comparative Examples of Calculating Information Content 30
   3.3 Two-Dimensional Plots 33
   3.4 Comparisons of Single Statistics 35
   References 40

4 **Discussion** 43
   References 44

Index 45

Chapter 1
Introduction

Characterizing the functional behavior of individual proteins in a variety of different contexts is an important step in understanding life at the molecular level. Endeavors such as understanding biological pathways, investigating disease, and developing drugs to cure those diseases depend on being able to describe the actions of individual proteins or genes, both in terms of their physiochemical molecular function, involvement in biological processes, and the subcellular location at which these actions are carried out.

Ontological representations have been widely used in biomedical sciences to standardize knowledge representation and exchange [58]. Modern ontologies are typically viewed as graphs in which vertices represent terms or concepts in the domain of interest and edges represent relational ties between terms (e.g., is-a, part-of). While, in theory, there are no restrictions on the types of graphs used to implement ontologies, hierarchical organizations, such as trees or directed acyclic graphs, have been frequently used in the systematization of biological experiments, organismal phenotypes, or structural and functional descriptions of biological macromolecules.

Several classification systems have been proposed to standardize annotation and to facilitate computation. With respect to defining a particular gene's phenotype, the Human Phenotype Ontology [59], the Unified Medical Language System [8] and the Disease Ontology [63] are human specific and predominantly constructed to address human disease. In the case of the Unified Medical Language System the focus is on defining associations between genes and medical disorders and, as a rare exception, terms are not hierarchically structured. Several ontologies also provide organism-independent terminology for defining phenotype such as the Vertebrate Skeletal Anatomy Ontology [13]. In terms of describing the bimolecular activity of a gene Enzyme Commission (EC) numbers [46] and the MIPS functional catalogue [61] are two well-accepted schemes; however, the most commonly used functional classification is the Gene Ontology (GO) [4].

The development of GO was based on the premise that the genomes of all living organisms are composed of evolutionarily related genes whose products perform functions derived from a finite molecular repertoire, limiting the lexicon necessary

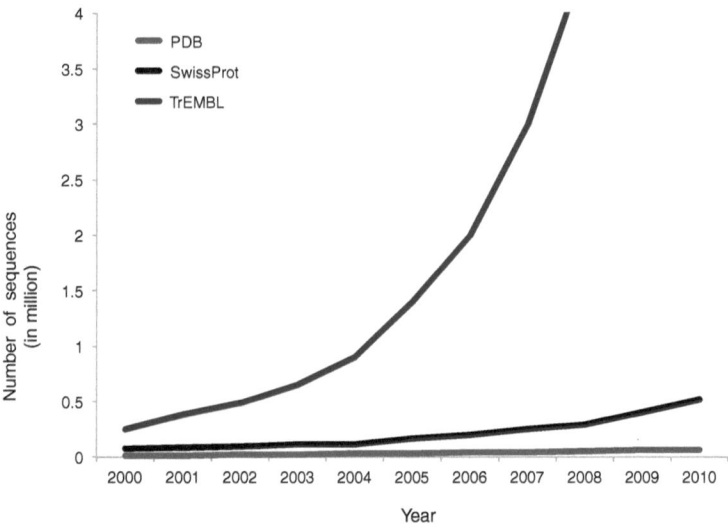

Fig. 1.1 The growth in the number of entries in three major databases is shown in millions of sequences. TrEMBL represents the compendium of all known sequences, regardless of how well characterized they are. PDB represents all sequences for which a structure has been experimentally determined. Swiss-Prot is a database of confirmed and relatively well-characterized proteins, although not all proteins have been assigned experimentally verified GO annotations

to describe them. GO provides three hierarchical classifications as directed acyclic graphs: molecular function ontology, biological process ontology, and cellular component. In addition to knowledge representation, GO has also facilitated large-scale analyses and automated annotation of gene product function [53].

In spite of the fact that there are increasingly more tools for the interrogation and description of protein behavior, there are still a large number of functionally uncharacterized proteins, with the gap between known sequences and experimentally characterized ones growing at an exponential pace (Fig. 1.1). Currently, there are about 50, 000 proteins with at least one experimentally annotated Gene Ontology (GO) term in Swiss-Prot [5]. However, owing to the numerous sequencing projects [40], the gap between annotated and nonannotated proteins has exceeded two orders of magnitude, and will only become wider. There is also a large amount of disparity in the distribution of experimental annotations among organisms, with most annotations occurring in model organisms (Fig. 1.2). Furthermore, while most experimentally annotated proteins come from model organisms, with the exception of yeast, less than half of the genome of any model organism has yet to be assigned experimentally characterized GO functions. The numbers in Fig. 1.2 do not illustrate the fact that many annotations are quite vague. Because of the multitude of ways a protein's function can be characterized, having a single annotation does not mean that protein has been fully annotated.

1 Introduction

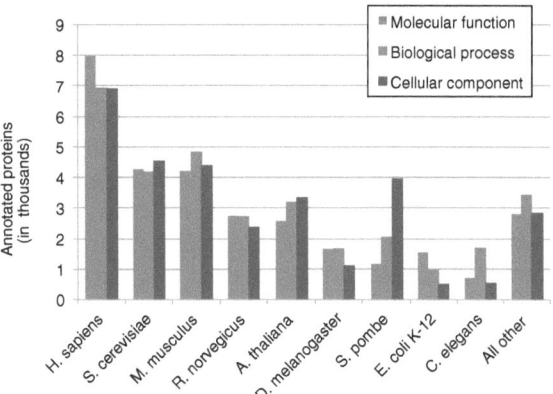

Fig. 1.2 The distribution of experimentally annotated (EXP, TAS, and IC evidence codes) proteins among well-characterized organisms and all other organisms for the three branches of the Gene Ontology. Annotations were taken from the January 2012 version of Swiss-Prot

As the rate of accumulation of uncharacterized sequences far outpaces the rate at which biological experiments can be carried out to characterize those sequences, computational function prediction has become increasingly useful for the global characterization of genomes and proteomes as well as for guiding biological experiments via prioritization [55, 64]. The growing importance of tools for the prediction ontological annotations, especially for proteins, presents the problem of how to accurately evaluate such tools. Because terms can automatically be associated with their ancestors in the GO graph, the task of an evaluation procedure can be framed as comparing the predicted graph with the graph represented by the true, experimentally verified annotations. It should be explicitly noted that the structure of the ontology introduces dependence between terms, dependencies that must be appropriately considered when comparing two graphs. This is obviously true for terms that share an ancestor/descendant relationship, but also for terms that are not related through direct ancestry. For example, although the terms "H3 histone acetyltransferase activity" and "H4 histone acetyltransferase activity" do not share a ancestor/descendant relationship and describe different behaviors of histone acetyltransferases, they nonetheless are not independent.

Protein function is also complex and context dependent. A single biological experiment rarely results in complete characterization of a protein's function. Furthermore, annotations might not always be meaningful; a point that is particularly evident in cases where only high-throughput experiments are used for functional characterization, leading to shallow annotation graphs. These types of annotations pose a problem in evaluation as the ground truth is incomplete and noisy.

Furthermore, different computational models produce different outputs that must be accounted for. Some models simply assign binary yes or no values to terms; while others might assign a different score to potentially each node in the ontology, with an expectation that a good decision threshold would be applied to provide useful annotations. Finally, a complicating factor is posed by the fact that GO, as most current ontologies, is generally unfinished and contains a range of specificities of functional descriptions at the same depth of the ontology [3].

When developing an evaluation metric it is important to keep in mind that, because both the experimental and predicted annotation of genes represent much smaller subgraphs of the larger GO graph, it is unlikely that a given computational method will provide an exact prediction of the experimental annotation. It is therefore necessary to develop metrics that facilitate calculating degrees of similarity between pairs of graphs and appropriately address dependency between nodes. Ideally, such a measure of similarity would be able to characterize both the level of correct prediction of the true (albeit incomplete) annotation but also the level of misannotation.

A final important factor related to the evaluation metric is its interpretability. Characterizing the predictor's performance should be meaningful to a downstream user. Ideally, an evaluation metric would have a simple probabilistic interpretation.

In this book, we develop an information-theoretic framework for evaluating the prediction accuracy of computer-generated ontological annotations. We first use the structure of the ontology to probabilistically model, via a Bayesian network, the prior distribution of protein experimental annotation. We then apply our metric to three protein function prediction algorithms selected to highlight the limitations of typically-considered evaluation metrics. We show that our metrics provide added value to the current analyses of the strengths and weaknesses of computational tools. Finally, we argue that our framework is probabilistically well founded and show that it can also be used to augment already existing evaluation metrics.

1.1 Background

The issue of performance evaluation is closely related to the problems of measuring similarity between pairs of graphs or sets. First, we note that a protein's annotation (experimental or predicted) is a graph containing a subset of nodes in the ontology together with edges connecting them. We use the term *leaf node* to describe a node that has no descendants in the annotation graph, although it is allowed to have descendants in the ontology. A set of leaf terms completely describes the annotation graph.

We roughly group both graph similarity and performance evaluation metrics into topological and probabilistic categories, and note that a particular metric may combine aspects from both. More elaborate distinctions are provided by Pesquita et al. [50] and Guzzi et al. [23]. Topological metrics rely on the structure of the ontology to perform evaluation and typically employ metrics that operate on sets of nodes and/or edges. A number of topological measures have been utilized, including the Jaccard and cosine similarity coefficients (the cosine function maps the binary term designations into a vector space), the shortest path-based distances [52], etc. In the context of classifier performance analysis, two common 2D metrics are the precision/recall curve and the Receiver Operating Characteristic (ROC) curve. Both curves are constructed based on the overlap in either edges or nodes between true and predicted terms and have been widely used to evaluate the performance of tools for the inference of GO annotations. They can also be used to provide a single statistic to rank

1.1 Background

classifiers through the maximum F-measure in the case of precision/recall curve or the area under the ROC curve. The area under the ROC curve has a limitation arising from the fact that the ontology is relatively large, but that the number of terms associated with a typical protein is relatively small. In practice, this results in specificities close to one regardless of the prediction, as long as the number of predicted terms is relatively small.

Although these statistics provide good feedback regarding multiple aspects of a predictor's performance, they do not always address node dependency or the problem of unequal specificity of functional annotations found at the same depth of the graph. Coupled with a large bias in the distribution of terms among proteins, prediction methods that simply learn the prior distribution of terms in the ontology could appear to have better performance than they actually do.

The second class of similarity/performance measures are probabilistic or information-theoretic metrics. Such measures assume an underlying probabilistic model over the ontology and use a database of proteins to learn the model. Similarity is then assessed by measuring the information content of the shared terms in the ontology but can also take into account the information content of the individual annotations. Unlike with topological measures where updates to the ontology affect similarity between objects, information-theoretic measures are also affected by changes in the underlying probabilistic model even if the structure of the ontology remains the same.

Probabilistic metrics closely follow and extend the methodology laid out by Resnik [56], which is based on the notion of information content between a pair of individual terms. These measures overcome biases related to the structure of the ontology; however, they have several drawbacks of their own. One that is especially important in the context of analyzing the performance of a predictor is that they only report a single statistic, namely the similarity or distance between two terms or sets of terms. This ignores the tradeoff between precision and recall that any predictor has to make. In the case of Resnik's metric, a prediction by any descendant of the true term will be scored as if it is an exact prediction. Similarly, a shallow prediction will be scored the same as a prediction that deviates from the true path at the same point, regardless of how deep the erroneous prediction might be. Although some of these weaknesses have been corrected in subsequent work [30, 39, 62], there remains the issue that the available probabilistic measures of semantic similarity resort to ad hoc solutions to address the common situation where proteins are annotated by graphs that contain multiple leaf terms [10]. Various approaches have been taken, including averaging between all pairs of leaf terms [41], finding the maximum among all pairs [57], or finding the best-match average, but each such solution lacks strong justification in general. For example, all-pair averaging (Eq. (2.19)) leads to anomalies where the exact prediction of an annotation containing a single leaf term u would be scored higher than the exact prediction of an annotation containing two distinct leaf terms u and v of equal information content, when it is more natural to think that the latter prediction should be scored higher (or that both should have perfect scores). Finally, all semantic similarity metrics that incorporate some form of pairwise matching between leaf terms tacitly assume that the objects to be compared are annotated by similar numbers of leaf terms. As such, they could produce undesirable solutions when applied to a wide range of prediction algorithms such as those outputting a very large number of predicted terms. For example, a predictor that predicts the whole ontology will obtain a perfect score when one applies the max-average method of averaging (Eq. (2.20)).

1.2 Protein Function Prediction Scenarios

In protein function prediction, there are two scenarios in which a computational model can be constructed: (i) given a new protein, the task of a classifier is to find all functional terms that the protein is associated with ("what is the function of this protein?"); and (ii) given a functional term, the task of a classifier is to find all proteins associated with this term ("what are the proteins associated with this function?"). The first case is generally referred to as an annotation problem [53], whereas the second is referred to as a gene prioritization problem [2, 71].

Undoubtedly, the two prediction scenarios are related because a perfect predictor of protein function would solve both questions at the same time. However, imperfect predictors may address one question better than the other. For example, a classifier built to address the first question is expected to be accurate in predicting all (or many) functional terms and that prediction scores over all functional terms are comparable. On the other hand, a predictor developed to consider only one functional term at a time need not consider any other term although it may be beneficial to do so. Such models can perform well even if they just rank all test proteins. In addition to these differences, the two types of models are evaluated in different ways (yet they use the same terminology).

Models that are concerned with predicting function on a previously unseen protein (scenario 1, above) need to devise evaluation metrics to estimate the expected accuracy of a predicted consistent graph P when the experimental (true) annotation of the protein is graph T. Alternatively, the models that are concerned with ranking the proteins according to their likelihood to be associated with a particular functional term v (scenario 2, above) need to be evaluated based on the expectation that a particular protein is associated with a functional term v. Here, the models are usually evaluated for each functional term v, one at a time. Evaluation metrics corresponding to the former problem are significantly more challenging than the metrics corresponding to the latter problem. In the latter case, one can simply consider a particular decision threshold for predicting whether a protein is associated with function v and then calculate the fraction of positive predictions that are correct (precision) as well as the fraction of proteins known to be associated with functional term v that have been retrieved (recall). Such evaluation has been discussed by Sharan et al. [64], among others. Calculating precision and recall for the former scenario is the topic of our study.

1.3 State of the Art Methods

Historically, sequence-based inference was the first strategy used to predict protein function, even if most studies at the time avoided explicitly relating homology and function [17]. Global and local sequence alignments were used to query sequence databases for similarities with a target protein. With the accumulation of experimentally determined protein functions, the most similar annotated sequences have traditionally been used to infer function [60]. Several methods have developed novel techniques for utilizing sequence alignment information as input to supervised learning methods [10, 34, 44, 66, 72]. More advanced methods exploited predicted physicochemical properties [12, 28, 29, 44], evolutionary relationships [6, 19, 20, 22, 42, 49], or the structure of functional ontologies in order to achieve different confidence levels for different ontological terms [7, 25, 43]. Microarrays [27], protein–protein interaction networks[15, 38, 45, 70], protein structures [26, 35, 47, 48] or a combination of data types [11, 33, 37, 66, 69] have also been exploited. However, most of these methods are limited to a few organisms where such data are available. One way or another, sequence alignment-based inference is the cornerstone of functional inference [24, 53].

Sequence alignment-based transfer of function has been thoroughly studied in the last decade, predominantly for enzymes [1, 16, 60, 67, 68, 73]. The results of these studies indicate that at least 60 % sequence identity, and more likely closer to 80 %, is required for the accurate transfer of the third level of EC classification. More sophisticated approaches were proposed as well: the GOtcha method was developed in order to take sequence alignment scores between a query protein and a functionally annotated database and overlay them on the functional ontology, cumulatively propagating such scores [43]. PFP refined this technique by incorporating alignments at very low significance levels and conditional probabilities that a protein is associated with pairs of functional terms [25]. Other methods such as ProtFun [28], ConFunc [72], GOsling [31], and GOstruct [66] were developed for high-throughput prediction tasks. Finally, phylogenetic methods attempt to exploit particular evolutionary relationships within a gene family [9, 18, 65]. Methods such as SIFTER [20] or ortholog identification methods [54] belong in this category. Several recent reviews provide good perspectives on protein function prediction at all scales [14, 21, 32, 35, 36, 51, 55, 60].

References

1. Addou, S., Rentzsch, R., Lee, D., Orengo, C.A.: Domain-based and family-specific sequence identity thresholds increase the levels of reliable protein function transfer. J. Mol. Biol. **387**(2), 416–430 (2009)
2. Aerts, S., Lambrechts, D., Maity, S., Van Loo, P., Coessens, B., De Smet, F., Tranchevent, L.-C., De Moor, B., Marynen, P., Hassan, B., et al.: Gene prioritization through genomic data fusion. Nat. Biotechnol. **24**(5), 537–544 (2006)

3. Alterovitz, G., Michael, X., Hill, D.P., Jane, L., Jonathan, L., Michael, C., Jonathan, D., Chris, M., Harris, M.A., Dolan, M.E., et al.: Ontology engineering. Nat. Biotechnol. **28**(2), 128–130 (2010)
4. Ashburner, M., et al.: Gene ontology: tool for the unification of biology. Nat. Genet. **25**(1), 25–29 (2000)
5. Amos, B., Rolf, A., Wu, C.H., Barker, W.C., Brigitte, B., Serenella, F., Elisabeth, G., Hongzhan, H., Rodrigo, L., Michele, M., et al.: The universal protein resource (UniProt). Nucleic Acids Res. **33**(1), D154–D159 (2005)
6. Bandyopadhyay, D., Huan, J., Liu, J., Prins, J., Snoeyink, J., Wang, W., Tropsha, A.: A structure-based function inference using protein family-specific fingerprints. Protein Sci. **15**(6), 1537–1543 (2006)
7. Barutcuoglu, Z., Schapire, R.E., Troyanskaya, O.G.: Hierarchical multi-label prediction of gene function. Bioinformatics **22**(7), 830–836 (2006)
8. Bodenreider, O.: The unified medical language system (UMLS): integrating biomedical terminology. Nucleic Acids Res. **32**(Database issue), D267–D270 (2004)
9. Brown, D., Sjolander, K.: Functional classification using phylogenomic inference. PLoS Comput. Biol. **2**(6), e77 (2006)
10. Clark, W.T., Radivojac, P.: Analysis of protein function and its prediction from amino acid sequence. Proteins Struct. Funct. Bioinf. **79**(7), 2086–2096 (2011)
11. Costello, J.C., Dalkilic, M.M., Beason, S.M., Gehlhausen, J.R., Patwardhan, R., Middha, S., Eads, B.D., Andrews, J.R., et al.: Gene networks in Drosophila melanogaster: integrating experimental data to predict gene function. Genome Biol. **10**(9), R97 (2009)
12. Cozzetto, D., Jones, D.T.: The contribution of intrinsic disorder prediction to the elucidation of protein function. Curr. Opin. Struct. Biol. **23**, 467–472 (2013)
13. Dahdul, W.M., Balhoff, J.P., Blackburn, D.C., Diehl, A.D., Haendel, M.A., Hall, B.K., Lapp, H., Lundberg, J.G., Mungall, C.J., Ringwald, M., et al.: A unified anatomy ontology of the vertebrate skeletal system. PLoS One **7**(12), e51070 (2012)
14. Dalkilic, M.M., Costello, J.C., Clark, W.T., Radivojac, P.: From protein-disease associations to disease informatics. Front. Biosci. **13**, 3391–3407 (2008)
15. Deng, M., Zhang, K., Mehta, S., Chen, T., Sun, F.: Prediction of protein function using protein-protein interaction data. J. Comput. Biol. **10**(6), 947–960 (2003)
16. Devos, D., Valencia, A.: Practical limits of function prediction. Proteins **41**(1), 98–107 (2000)
17. Doolittle, R.F.: Of URFS and ORFS: a primer on how to analyze derived amino acid sequences. University Science Books, Mill Valley (1986)
18. Eisen, J.A.: Phylogenomics: improving functional predictions for uncharacterized genes by evolutionary analysis. Genome Res. **8**, 163–167 (1998)
19. Enault, F., Suhre, K., Claverie, J.-M.: Phydbac "Gene Function Predictor": a gene annotation tool based on genomic context analusis. BMC Bioinf. **6**(257), 247 (2005)
20. Engelhardt, B.E., Jordan, M.I., Muratore, K.E., Brenner, S.E.: Protein molecular function prediction by Bayesian phylogenomics. PLoS Comput. Biol. **1**(5), e45 (2005)
21. Friedberg, I.: Automated protein function prediction-the genomic challenge. Briefings Bioinf. **7**(3), 225–242 (2006)
22. Gaudet, P., Livstone, M.S., Lewis, S.E., Thomas, P.D.: Phylogenetic-based propagation of functional annotations within the gene ontology consortium. Briefings Bioinf. **12**(5), 449–462 (2011)
23. Guzzi, P.H., et al.: Semantic similarity analysis of protein data: assessment with biological features and issues. Briefings Bioinf. **13**(5), 569–585 (2012)
24. Hamp, T., Kassner, R., Seemayer, S., Vicedo, E., Schaefer, C., Achten, D., Auer, F., Boehm, A., Braun, T., Hecht, M., et al.: Homology-based inference sets the bar high for protein function prediction. BMC Bioinf. **14**(Suppl 3), S7 (2013)
25. Hawkins, T., Luban, S., Kihara, D.: Enhanced automated function prediction using distantly related sequences and contextual association by PFP. Protein Sci. **15**(6), 1550–1556 (2006)
26. Hermann, J.C., Marti-Arbona, R., Fedorov, A.A., Fedorov, E., Almo, S.C., Shoichet, B.K., Raushel, F.M.: Structure-based activity prediction for an enzyme of unknown function. Nature **448**(7155), 775–779 (2007)

References

27. Huttenhower, C., Hibbs, M., Myers, C., Troyanskaya, O.G.: A scalable method for integration and functional analysis of multiple microarray datasets. Bioinformatics **22**(23), 2890–2897 (2006)
28. Jensen, L.J., Gupta, R., Staerfeldt, H.H., Brunak, S.: Prediction of human protein function according to gene ontology categories. Bioinformatics **19**(5), 635–642 (2003)
29. Jensen, L.J., Gupta, R., Blom, N., Devos, D., Tamames, J., Kesmir, C., Nielsen, H., Stærfeldt, H.H., Rapacki, K., Workman, C., et al.: Prediction of human protein function from post-translational modifications and localization features. J. Mol. Biol. **319**(5), 1257–1266 (2002)
30. Jiang, J.J., Conrath, D.W.: Semantic similarity based on corpus statistics and lexical taxonomy. In: International Conference on Research in Computational Linguistics, pp. 19–33 (1997)
31. Jones, C.E., Schwerdt, J., Bretag, T.A., Baumann, U., Brown, A.L.: Gosling: a rule-based protein annotator using blast and go. Bioinformatics **24**(22), 2628–2629 (2008)
32. Kann, M.G.: Protein interactions and disease: computational approaches to uncover the etiology of diseases. Briefings Bioinf. **8**(5), 333–346 (2007)
33. Kourmpetis, Y.A.I., van Dijk, A.D.J., Bink, M.C.A.M., van Ham, R.C.H.J., Ter Braak, C.J.F.: Bayesian Markov random field analysis for protein function prediction based on network data. PloS One **5**(2), e9293 (2010)
34. Kourmpetis, Y.A.I., van Dijk, A.D.J., ter Braak, C.J.F.: Gene ontology consistent protein function prediction: the falcon algorithm applied to six eukaryotic genomes. Algorithms Mol. Biol. **8**(1), 10 (2013)
35. Laskowski, R.A., Thornton, J.M.: Understanding the molecular machinery of genetics through 3D structures. Nat. Rev. Genet. **9**(2), 141–151 (2008)
36. Lee, D., Redfern, O., Orengo, C.: Predicting protein function from sequence and structure. Nat. Rev. Mol. Cell Biol. **8**(12), 995–1005 (2007)
37. Lee, I., Date, S.V., Adai, A.T., Marcotte, E.M.: A probabilistic functional network of yeast genes. Science **306**(5701), 1555–1558 (2004)
38. Letovsky, S., Kasif, S.: Predicting protein function from protein/protein interaction data: a probabilistic approach. Bioinformatics **19**(Suppl 1), i197–204 (2003)
39. Lin, D.: An information-theoretic definition of similarity. In: Proceedings of the 15th International Conference on Machine Learning, pp. 296–304. Morgan Kaufmann, San Francisco (1998)
40. Liolios, K., Mavromatis, K., Tavernarakis, N., Kyrpides, N.C.: The genomes on line database (gold) in 2007: status of genomic and metagenomic projects and their associated metadata. Nucleic Acids Res. **36**(Database issue), D475–D479 (2008)
41. Lord, P.W., et al.: Investigating semantic similarity measures across the gene ontology: the relationship between sequence and annotation. Bioinformatics **19**(10), 1275–1283 (2003)
42. Marcotte, E.M., Pellegrini, M., Ng, H.-L., Rice, D.W., Yeates, T.O., Eisenberg, D.: Detecting protein function and protein-protein interactions from genome sequences. Science **285**(5428), 751–753 (1999)
43. Martin, D.M., et al.: GOtcha: a new method for prediction of protein function assessed by the annotation of seven genomes. BMC Bioinf. **5**, 178 (2004)
44. Minneci, F., Piovesan, D., Cozzetto, D., Jones, D.T.: FFPred 2.0: improved homology-independent prediction of gene ontology terms for eukaryotic protein sequences. PLoS One **8**(5), e63754 (2013)
45. Nabieva, E., Jim, K., Agarwal, A., Chazelle, B., Singh, M.: Whole-proteome prediction of protein function via graph-theoretic analysis of interaction maps. Bioinformatics **21**(Suppl 1), i302–310 (2005)
46. (NC-IUBMB) NCotIUoBaMB. Enzyme nomenclature. Academic Press, New York (1992)
47. Pal, D., Eisenberg, D.: Inference of protein function from protein structure. Structure **13**(1), 121–130 (2005)
48. Pazos, F., Sternberg, M.J.: Automated prediction of protein function and detection of functional sites from structure. Proc. Nat. Acad. Sci. U.S.A. **101**(41), 14754–14759 (2004)
49. Pellegrini, M., Marcotte, E.M., Thompson, M.J., Eisenberg, D., Yeates, T.O.: Assigning protein functions by comparative genome analysis: protein phylogenetic profiles. Proc. Nat. Acad. Sci. U.S.A. **96**(8), 4285–4288 (1999)

50. Pesquita, C., et al.: Semantic similarity in biomedical ontologies. PLoS Comput. Biol. **5**(7), e1000443 (2009)
51. Punta, M., Ofran, Y.: The rough guide to in silico function prediction, or how to use sequence and structure information to predict protein function. PLoS Comput. Biol. **4**(10), e1000160 (2008)
52. Rada, R., et al.: Development and application of a metric on semantic nets. IEEE Trans. Syst. Man Cybern. **19**(1), 17–30 (1989)
53. Radivojac, P., Clark, W.T., et al.: A large-scale evaluation of computational protein function prediction. Nat. Methods **10**(3), 221–227 (2013)
54. Remm, M., Storm, C.E., Sonnhammer, E.L.: Automatic clustering of orthologs and in-paralogs from pairwise species comparisons. J. Mol. Biol. **314**(5), 1041–1052 (2001)
55. Rentzsch, R., Orengo, C.A.: Protein function prediction-the power of multiplicity. Trends Biotechnol. **27**(4), 210–219 (2009)
56. Resnik, P.: Using information content to evaluate semantic similarity in a taxonomy. In: Proceedings of the 14th International Joint Conference on Artificial Intelligence, pp. 448–453 (1995)
57. Resnik, P.: Semantic similarity in a taxonomy: an information-based measure and its application to problems of ambiguity in natural language. J. Artif. Intell. Res. **11**, 95–130 (1999)
58. Robinson, P.N., Bauer, S.: Introduction to bio-ontologies. CRC Press, Boca Raton (2011)
59. Robinson, P.N., Mundlos, S.: The human phenotype ontology. Clin. Genetics **77**(6), 525–534 (2010)
60. Rost, B., Liu, J., Nair, R., Wrzeszczynski, K.O., Ofran, Y.: Automatic prediction of protein function. Cell. Mol. Life Sci. **60**(12), 2637–2650 (2003)
61. Ruepp, A., Zollner, A., Maier, D., Albermann, K., Hani, J., Mokrejs, M., Tetko, I., Guldener, U., Mannhaupt, G., Munsterkotter, M., Mewes, H.W.: The FunCat, a functional annotation scheme for systematic classification of proteins from whole genomes. Nucleic Acids Res. **32**(18), 5539–5545 (2004)
62. Schlicker, A., et al.: A new measure for functional similarity of gene products based on gene ontology. BMC Bioinf. **7**, 302 (2006)
63. Schriml, L.M., Arze, C., Nadendla, S., Chang, Y.-W.W., Mazaitis, M., Felix, V., Feng, G., Kibbe, W.A.: Disease ontology: a backbone for disease semantic integration. Nucleic Acids Res. **40**(D1), D940–D946 (2012)
64. Sharan, R., et al.: Network-based prediction of protein function. Mol. Syst. Biol. **3**, 88 (2007)
65. Škunca, N., Bošnjak, M., Kriško, A., Panov, P., Džeroski, S., Šmuc, T., Supek, F.: Phyletic profiling with cliques of orthologs is enhanced by signatures of paralogy relationships. PLoS Comput. Biol. **1553**, 734X (2013)
66. Sokolov, A., Ben-Hur, A.: Hierarchical classification of gene ontology terms using the Gostruct method. J. Bioinf. Comput. Biol. **8**(2), 357–376 (2010)
67. Tian, W., Skolnick, J.: How well is enzyme function conserved as a function of pairwise sequence identity? J. Mol. Biol. **333**(4), 863–882 (2003)
68. Todd, A.E., Orengo, C.A., Thornton, J.M.: Evolution of function in protein superfamilies, from a structural perspective. J. Mol. Biol. **307**(4), 1113–1143 (2001)
69. Troyanskaya, O.G., Dolinski, K., Owen, A.B., Altman, R.B., Botstein, D.: A bayesian framework for combining heterogeneous data sources for gene function prediction (in Saccharomyces cerevisiae). Proc. Nat. Acad. Sci. U.S.A. **100**(14), 8348–8353 (2003)
70. Vazquez, A., Flammini, A., Maritan, A., Vespignani, A.: Global protein function prediction from protein-protein interaction networks. Nat. Biotechnol. **21**(6), 697–700 (2003)
71. Warde-Farley, D., Donaldson, S.L., Comes, O., Zuberi, K., Badrawi, R., Chao, P., Franz, M., Grouios, C., Kazi, F., Lopes, CT., et al.: The genemania prediction server: biological network integration for gene prioritization and predicting gene function. Nucleic Acids Res. **38** (suppl 2), W214–W220 (2010)

72. Wass, M.N., Sternberg, M.J.: ConFunc-functional annotation in the twilight zone. Bioinformatics **24**(6), 798–806 (2008)
73. Wilson, C.A., Kreychman, J., Gerstein, M.: Assessing annotation transfer for genomics: quantifying the relations between protein sequence, structure and function through traditional and probabilistic scores. J. Mol. Biol. **297**(1), 233–249 (2000)

Chapter 2
Methods

We previously introduced information-theoretic metrics for evaluating classification performance in protein function prediction which we describe here [2]. In this learning scenario, the input space $\mathcal{X}$ represents proteins, whereas the output space $\mathcal{Y}$ contains directed acyclic graphs describing protein function according to the Gene Ontology (GO).

Because of the hierarchical nature of GO, both experimental and computational annotations need to satisfy the *consistency requirements*:

i. If vertex (term) v from the ontology is true, then all of its ancestors must also be true.
ii. If vertex (term) v from the ontology is false, then all of its descendants must also be false.

By enforcing these requirements, we frame the task of a classifier as assigning the best consistent subgraph of the ontology to each new protein and output a prediction score for this subgraph and/or each predicted term.

We simplify the exposition by referring to such graphs as prediction or annotation graphs. In addition, we frequently treat consistent graphs as sets of nodes or functional terms and use set operations to manipulate them.

We now proceed to provide a definition for the information content of a (consistent) subgraph in the ontology. Then, using this definition, we derive information-theoretic performance evaluation metrics for comparing pairs of graphs.

2.1 Calculating the Joint Probability of a Graph

Let each term in the ontology be a binary random variable and consider a fixed but unknown probability distribution over $\mathcal{X}$ and $\mathcal{Y}$ according to which the quality of a prediction process will be evaluated. We shall assume that the prior distribution of a target can be factorized according to the structure of the ontology, i.e., we assume a Bayesian network as the underlying data generating process for the target variable. According to this assumption, each term is independent of its ancestors given its parents and, thus, the full Joint probability can be factorized as a product of individual terms obtained from the set of Conditional probability tables associated with each term [7],

$$\Pr(\mathbf{V}) = \prod_{v \in \mathbf{V}} \Pr(v|\mathcal{P}(v)), \tag{2.1}$$

where $\mathbf{V}$ denotes all vertices the ontology, v denotes a vertex in $\mathbf{V}$ and $\mathcal{P}(v)$ is the set of parent nodes of v. Here we use $\mathbf{V}$ (all terms) instead of T (only true terms) to illustrate the fact that we are calculating the probability of a given configuration of the full ontology with the inclusion of all true and false terms. In practice the true terms, or the subgraph T, are generally all that is considered both in the context of this chapter and in other papers addressing similar topics.

The scope of terms considered can be reduced when calculating the joint probability of a configuration of $\mathbf{V}$ without affecting the final probability value. Because of the enforced consistency requirement (i.e., all ancestors of true terms are true; all descendants of false terms are false) the full joint probability of a configuration of the ontology, $\mathbf{V}$, can be calculated by considering only terms whose parents are all true. Equation 2.1 can be rewritten as

$$\Pr(\mathbf{V}) = \prod_{v \in T \cup \mathcal{C}(T)} \Pr(v|\mathcal{P}(v)), \tag{2.2}$$

where T denotes all terms in $\mathbf{V}$ that are true, $\mathcal{C}(T)$ defines false terms all of whose parents are true (children of leaf terms in T), and $T \cup \mathcal{C}(T)$ defines the union of the two sets.

The derivation of Eq. 2.2 can be obtained by considering Fig. 2.1 which illustrates a sample ontology with 9 terms and Table 2.1 which represents the conditional probability distribution, or table, for vertex g. If we write the joint probability of $\mathbf{V}$ using Eq. 2.1 we obtain

$$\begin{aligned}\Pr(\mathbf{V}) = {} & \Pr(a=1) \times \Pr(b=0|a=1) \times \Pr(c=1|a=1) \times \Pr(d=0|b=0) \\ & \times \Pr(e=1|c=1) \times \Pr(f=0|c=1) \times \Pr(g=0|d=0, e=1) \\ & \times \Pr(h=0|f=0) \times \Pr(i=0|f=0).\end{aligned}$$

First, we see that terms h and i contribute nothing to the joint probability because the probability a term is false given its parents are all false is always equal to 1. Although b and f are false, because all of their parents are true the resulting probability is not 1. Term g can also be ignored because one of its parents is false as illustrated by considering Table 2.1. The resulting joint probability can subsequently be rewritten as

$$\begin{aligned}\Pr(\mathbf{V}) = {} & \Pr(a=1) \times \Pr(b=0|a=1) \times \Pr(c=1|a=1) \\ & \times \Pr(e=1|c=1) \times \Pr(f=0|c=1).\end{aligned}$$

2.1 Calculating the Joint Probability of a Graph

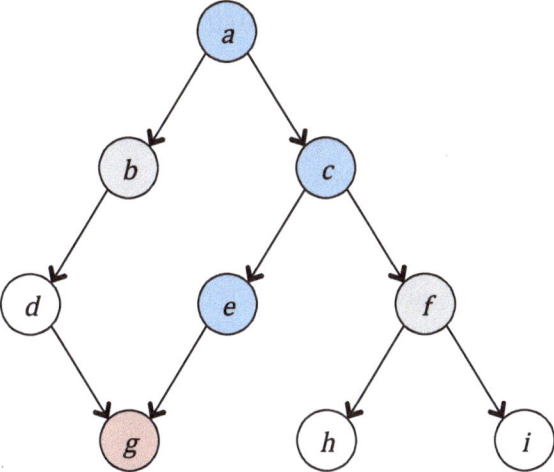

Fig. 2.1 A directed acyclic graph representing relationships between terms in a potential ontology. True terms, or T, are represented by *blue* colored vertices: $\{a, c, e\}$. The children of true terms, $C(T)$, are denoted by *grey* and *light-red* vertices: $\{b, f, g\}$. Terms which do not contribute anything to the full joint probability are shown in *white* and *light-red*: $\{d, g, h, i\}$. Term g is colored *light red* to point out a special case when calculating the full joint probability of a graph. Although one of term g's parents are true, because it has one false parent it will not contribute to the calculation of the full joint probability because the probability it is false is always equal to one (Table 2.1). All *blue* and *grey* terms taken together comprise the *markov blanket* of T, and are necessary when calculating the full joint probability of a configuration, or a combination of true and false values for all terms, in **V**

Table 2.1 A table showing a conditional probability distribution for term g from Fig. 2.1

d	e	Pr(g) True	False
True	True	.7	.3
True	False	0	1
False	True	0	1
False	False	0	1

The purpose of this table is to illustrate how the consistency requirement affects joint probability tables in the Bayesian network. If at least one of a term's parents is false it is guaranteed that that term is also false and contributes nothing. This is shown by considering the last three rows in the table

Because in practice most annotations only draw upon a small fraction of terms, the vast majority of terms will be negative and will not contribute to the calculation of a given joint probability. For this reason, although calculating the full joint probability has an asymptotic upper bound defined by the number of terms in the ontology, in practice it should have much lower complexity.

In this context, we are only interested in marginal probabilities that a protein is experimentally associated with a consistent subgraph T in the ontology. This probability can be expressed as

$$\Pr(T) = \prod_{v \in T} \Pr(v|\mathcal{P}(v)). \qquad (2.3)$$

Equation 2.3 can be derived from the full joint factorization by first marginalizing over the leaves of the ontology and then moving toward the root(s) for all nodes not in T.

Although marginalizing over all negative terms gives a formulation that deviates from calculating the full joint probability it both simplifies calculating the information content of a subgraph and can be philosophically justified. Unobserved annotations are only putative negative observations because explicit negative annotations are rarely deposited in biological databases and in ignoring them we are implicitly recognizing them as such.

2.1.1 Calculating the Information Content of a Graph

Now that we have properly defined the joint probability of a set of terms, or subgraph of the ontology, calculating the information content of those terms is relatively straightforward. The information content of a subgraph can be thought of as the number of bits of information one would receive about a protein if it were annotated with that particular subgraph. We calculate the information content of a subgraph T in a straightforward manner as

$$i(T) = \log \frac{1}{\Pr(T)}$$

and use a base 2 logarithm as a matter of convention. The information content of a subgraph T can now be expressed by combining the previous two equations as

$$i(T) = \sum_{v \in T} \log \frac{1}{\Pr(v|\mathcal{P}(v))} \qquad (2.4a)$$

$$= \sum_{v \in T} ia(v), \qquad (2.4b)$$

where, to simplify the notation, we use $ia(v)$ to represent the negative logarithm of $\Pr(v|\mathcal{P}(v))$. Term $ia(v)$ can be thought of as the increase, or accretion, of information obtained by adding a child term to a parent term, or set of parent terms, in an annotation. We will refer to $ia(v)$ as *information accretion* (perhaps information gain would be a better term, but because it is frequently used in other applications to describe an expected reduction in entropy, we avoid it in this situation).

A simple ontology containing five terms together with a conditional probability table associated with each node is shown in Fig. 2.2a. Note that because of the graph consistency requirement, each conditional probability table is limited to a single

2.1 Calculating the Joint Probability of a Graph

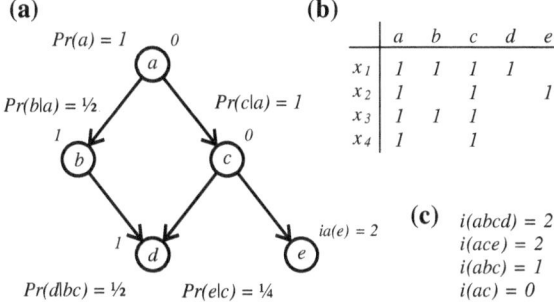

Fig. 2.2 An example of an ontology, data set, and calculation of information content. **a** An ontology viewed as a Bayesian network together with a conditional probability table assigned to each node. Each conditional probability table is limited to a single number due to the consistency requirement in assignments of protein function. Information accretion calculated for each node, e.g., $ia(e) = -\log \Pr(e|c) = 2$, are shown in *grey* next to each node. **b** A data set containing four proteins whose functional annotations are generated according to the probability distribution from the Bayesian network. **c** The total information content associated with each protein found in panel **b**; e.g., $i(ace) = ia(a) + ia(c) + ia(e) = 2$. Note that $i(ab) = 1$ and $i(abcde) = 4$ although proteins with such annotation have not been observed in part **b**

number. For example, at node b in the graph, the probability $\Pr(b = 1|a = 1)$ is the only one necessary because $\Pr(b = 0|a = 1) = 1 - \Pr(b = 1|a = 1)$ and because $\Pr(b = 1|a = 0)$ is guaranteed to be 0. In Fig. 2.2b we show a sample data set of four proteins functionally annotated according to the distribution defined by the Bayesian network. In Fig. 2.2c, we show the total information content for each of the four annotation graphs.

2.1.2 Comparing Two Annotation Graphs

We now consider a situation in which a protein's true and predicted function are represented by graphs T and P, respectively. We define two metrics that can be thought of as the information-theoretic analogs of recall and precision, and refer to them as remaining uncertainty and misinformation, respectively.

Definition 1. The *remaining uncertainty* about the protein's true annotation corresponds to the information about the protein that is not yet provided by the graph P. More formally, we express the remaining uncertainty (ru) as

$$ru(T, P) = \sum_{v \in T \setminus P} ia(v), \qquad (2.5)$$

which is simply the total information content of the nodes in the ontology that are contained in true annotation T but not in the predicted annotation P. Note that, in a slight abuse of notation, we apply set operations to graphs to manipulate only the vertices of these graphs.

Definition 2. The *misinformation* introduced by the classifier corresponds to the total information content of the nodes along incorrect paths in the prediction graph P. More formally, the misinformation is expressed as

$$mi(T, P) = \sum_{v \in P \setminus T} ia(v), \qquad (2.6)$$

which quantifies how misleading a predicted annotation is.

Here, a perfect prediction (one that achieves $P = T$) leads to $ru(T, P) = mi(T, P) = 0$. However, both $ru(T, P)$ and $mi(T, P)$ can be infinite in the limit. In practice, though, $ru(T, P)$ is bounded by the information content of the particular annotation, while $mi(T, P)$ is only limited by the number of annotations a predictor chooses to return.

To illustrate calculation of remaining uncertainty and misinformation, in Fig. 2.2 we show a sample ontology where the true annotation of a protein T is determined by the two leaf terms t_1 and t_2, whereas the predicted subgraph P is determined by the leaf terms p_1 and p_2. The remaining uncertainty $ru(T, P)$ and misinformation $mi(T, P)$ can now be calculated by adding the information accretion corresponding to the nodes circled in grey.

Finally, this framework can be used to define the semantic similarity between the protein's true annotation and the predicted annotation without relying on identifying an individual common ancestor between pairs of leaves (this node is usually referred to as the most informative common ancestor; [4]) (Fig. 2.3). The information content of the subgraph shared by T and P is one such possibility; i.e.,

$$s(T, P) = \sum_{v \in T \cap P} ia(v).$$

2.1.3 Measuring the Quality of Function Prediction

A typical predictor of protein function usually outputs scores that indicate the strength (e.g., posterior probabilities) of predictions for each term in the ontology. To address this situation, the concepts of remaining uncertainty and misinformation need to be considered as a decision threshold function of a τ. In such a scenario, predictions with scores greater than or equal to τ are considered positive predictions, while the remaining associations are considered negative (if the strength of a prediction is expressed via P-values or e-values, values lower than the threshold would indicate positive predictions). Regardless of the situation, every decision threshold results in a separate pair of values corresponding to the remaining uncertainty $ru(T, P(\tau))$ and misinformation $mi(T, P(\tau))$.

2.1 Calculating the Joint Probability of a Graph

$$s(T, P) = \sum_{v \in T \cap P} ia(v).$$

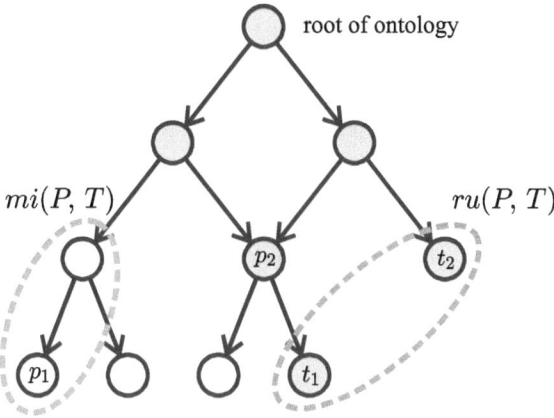

Fig. 2.3 Illustration of calculating remaining uncertainty and misinformation given a predicted annotation graph P and a graph of true annotations T. Graphs P and T are uniquely determined by the leaf nodes p_1, p_2, t_1, and t_2, respectively. Nodes colored in *grey* represent graph T. Nodes circled in *grey* are used to determine remaining uncertainty (ru; *right side*) and misinformation (mi; *left side*) between T and P

The remaining uncertainty and misinformation for a previously unseen protein can be calculated as expectations over the data generating probability distribution. Practically, this can be performed by averaging over the entire set of proteins used in evaluation, i.e.,

$$ru(\tau) = \frac{1}{n} \sum_{i=1}^{n} ru(T_i, P_i(\tau)) \qquad (2.7)$$

and

$$mi(\tau) = \frac{1}{n} \sum_{i=1}^{n} mi(T_i, P_i(\tau)) \qquad (2.8)$$

where n is the number of proteins in the data set, T_i is the true set of terms for protein x_i, and $P_i(\tau)$ is the set of predicted terms for protein x_i given decision threshold τ. Note that once the set of terms with scores greater than or equal to τ is determined, the set $P_i(\tau)$ is composed of the unique union of the ancestors of all predicted terms. As the decision threshold is moved from its minimum to its maximum value, the pairs of $(ru(\tau), mi(\tau))$ will result in a curve in 2D space. We refer to such a curve using $(ru(\tau), mi(\tau))_\tau$. Removing the normalizing constant ($\frac{1}{n}$) from the above equations would result in the total remaining uncertainty and misinformation associated with a database of proteins and a set of predictions.

2.1.4 Weighted Metrics

One disadvantage of definitions in Eqs. 2.7 and 2.8 is that an equal weight is given to proteins with low and high information content annotations when averaging. To address this, we assign a weight to each protein according to the information content of its experimental annotation. This formulation naturally downweights proteins with less informative annotations compared to proteins with rare, and therefore more informative (surprising), annotations. In biological data sets, frequently seen annotations have a tendency to be incomplete or shallow annotation graphs and arise due to the limitations or high-throughput nature of some experimental protocols. We define *weighted remaining uncertainty* as

$$wru(\tau) = \frac{\sum_{i=1}^{n} i(T_i) \cdot ru(T_i, P_i(\tau))}{\sum_{i=1}^{n} i(T_i)} \tag{2.9}$$

and *weighted misinformation* as

$$wmi(\tau) = \frac{\sum_{i=1}^{n} i(T_i) \cdot mi(T_i, P_i(\tau))}{\sum_{i=1}^{n} i(T_i)} \tag{2.10}$$

2.1.5 Semantic Distance

Finally, to provide a single performance measure which can be used to rank and evaluate protein function prediction algorithms, we introduce *semantic distance* as the minimum distance from the origin to the curve $(ru(\tau), mi(\tau))_\tau$. More formally, the semantic distance S_k is defined as

$$S_k = \min_{\tau} (ru^k(\tau) + mi^k(\tau))^{\frac{1}{k}}, \tag{2.11}$$

where k is a real number ≥ 1. Setting $k = 2$ results in the minimum Euclidean distance from the origin. The preference for Euclidean distance ($k = 2$) over say Manhattan distance ($k = 1$) is to penalize unbalanced predictions with respect to the depth of predicted and experimental annotations.

2.1.6 Precision and Recall

In order to contrast the semantic distance-based evaluation with more conventional performance measures, in this section we briefly introduce precision and recall for measuring functional similarity. As before, we consider a set of propagated experimental terms T and predicted terms $P(\tau)$, and define precision as the fraction of terms predicted correctly. More specifically,

$$pr(T, P(\tau)) = \frac{|T \cap P(\tau)|}{|P(\tau)|},$$

where $|\cdot|$ is the set cardinality operator. Only proteins for which the prediction set is nonempty can be used to calculate average precision. To address this issue the root term is counted as a prediction for all proteins. Similarly, recall is defined as the fraction of experimental (true) terms which were correctly predicted, i.e.,

$$rc(T, P(\tau)) = \frac{|T \cap P(\tau)|}{|T|}.$$

As before, precision $pr(\tau)$ and recall $rc(\tau)$ for the entire data set are calculated as averages over the entire set of proteins (note that an alternative definition of precision and recall given by [13] is described in Sect. 2.1.8). Finally, to provide a single evaluation measure we use the maximum F-measure over all decision thresholds. For a particular set of terms T and $P(\tau)$, F-measure is calculated as the Harmonic mean of precision and recall. More formally, the final evaluation metric is calculated as

$$F_{\max} = \max_{\tau} \left\{ 2 \cdot \frac{pr(\tau) \cdot rc(\tau)}{pr(\tau) + rc(\tau)} \right\} \quad (2.12)$$

where $pr(\tau)$ and $rc(\tau)$ are calculated by averaging over the data set.

2.1.6.1 Information-Theoretic Weighted Formulation

The definition of information accretion and the use of a probabilistic framework defined by the Bayesian network enable the straightforward application of information accretion to weight each term in the ontology. Therefore, it is easy to generalize the definitions of precision and recall from the previous section into a weighted formulation. Here, weighted precision and weighted recall can be expressed as

$$wpr(T, P(\tau)) = \frac{\sum_{v \in T \cap P(\tau)} ia(v)}{\sum_{v \in P(\tau)} ia(v)}$$

and

$$wrc(T, P(\tau)) = \frac{\sum_{v \in T \cap P(\tau)} ia(v)}{\sum_{v \in T} ia(v)}.$$

Weighted precision $wpr(\tau)$ and recall $wrc(\tau)$ can then be calculated as weighted averages over the database of proteins, as in Eqs. 2.9 and 2.10.

In addition to weighted precision and recall our framework also facilitates calculating weighted specificity

$$wsp(T, P(\tau)) = \frac{\sum_{v \in T^c \cap P^c(\tau)} ia(v)}{\sum_{v \in T^c} ia(v)}$$

where T^c represents the complement of set T ($G \setminus T$).

2.1.7 Supplementary Evaluation Metrics

When calculating remaining uncertainty, misinformation, precision, and recall only consistent subgraphs of the Gene Ontology were considered. Under this framework, if a protein is annotated with multiple terms (either experimentally determined or predicted), as in Fig. 2.2, we determine consistent graphs T (true) or P (predicted) by recursively propagating annotations toward the root(s) of the ontology and taking a union of all terms encountered along the way. In each of these measures, it is sufficient to only consider vertices (terms) in the annotation graphs and calculate the similarity measure by manipulating vertices in an additive fashion. For example, each vertex in T or P counts equally in the precision/recall-based evaluation while the information accretion is used to weigh the vertices in the ru–mi-based evaluation.

A distinctly different approach can be taken by considering, on an individual basis, each leaf term that comprises a set T or P. This is the approach taken to calculate various information-theoretic metrics [6, 8, 9, 11, 12] as well as to provide an alternative definition of precision and recall [13]. In this context the sets of leaf terms that define T and P (which we refer to as $\mathcal{L}(T)$ and $\mathcal{L}(P)$ respectively, and formally introduce below) are used to calculate a given metric. After calculating all pairwise metrics between the leaf terms, several different methods for averaging these scores can be applied to create a single similarity (or distance) value between T and P. We discuss these approaches below.

2.1 Calculating the Joint Probability of a Graph

2.1.7.1 Basic Definitions

Suppose we are given an ontology in the form of directed acyclic graph $G = (V, E)$, where V is a set of vertices and $E \subset V \times V$ is the set of edges. In this graph, given an edge $(u, v) \in E$, we refer to vertex u as a parent of v and, alternatively, to vertex v as a child of u. We also consider a set of all ancestors of v, $\mathcal{A}(v)$, and find this set by recursively identifying parents of all discovered nodes starting with v until the root(s) of the ontology is (are) reached. For mathematical convenience, we consider vertex v to be a member of $\mathcal{A}(v)$. Finally, given two vertices u and v, we define a set of common ancestor nodes between these two vertices as $\mathcal{A}(u, v)$. Thus, $\mathcal{A}(u, v) = \mathcal{A}(u) \cap \mathcal{A}(v)$.

Consider now a consistent annotation graph T, where the set of vertices in T is a subset of vertices in G. We define $\mathcal{L}(T)$, or the set of Leaf terms represented by T, as

$$\mathcal{L}(T) = \{u : u \in T \wedge \neg \exists ((u, v) \in E \wedge v \in T)\}. \tag{2.13}$$

In other words, $\mathcal{L}(T)$ contains only those vertices (terms) from T that do not have children in T. Thus, the leaf terms are defined with respect to a particular annotation graph T and generally differ from the leaf nodes in the ontology.

2.1.7.2 Information-Theoretic Metrics Between Pairs of Vertices

When calculating the information-theoretic metrics of [6, 8, 9, 11, 12], we calculate the information content of an individual term $v \in V$ as

$$i(v) = \log \frac{1}{\Pr(v)} \tag{2.14}$$

where $\Pr(v)$ can be calculated as the relative frequency of term v among experimentally annotated proteins. The semantic similarity between two distinct terms u and v as defined by [11] was calculated as

$$s_R(u, v) = \max_{w \in \mathcal{A}(u,v)} \{i(w)\}, \tag{2.15}$$

where $\mathcal{A}(u, v)$, as mentioned above, defines the set of common ancestors of terms u and v. Similarity as defined by Lin [8] was calculated as

$$s(u, v) = \frac{s_R(u, v)}{i(u) + i(v)}, \tag{2.16}$$

and as defined by Schlicker et al. [12] as

$$s(u, v) = \frac{s_R(u, v)}{i(u) + i(v)} \cdot (1 - \min_{w \in \mathcal{A}(u,v)} \{\Pr(w)\}). \tag{2.17}$$

Finally, the distance metric defined by [6] was calculated as

$$d(u, v) = i(u) + i(v) - 2 \cdot s_R(u, v). \tag{2.18}$$

2.1.7.3 Information-Theoretic Metrics Between Pairs of Graphs

Since the above metrics are only defined for two distinct terms, it is necessary to provide a mechanism to utilize these metrics in instances where a protein is annotated with graphs containing multiple leaf terms. Given two nonempty consistent annotation graphs of true and predicted terms, T and P, and the set of leaf terms that define each set, $\mathcal{L}(T)$ and $\mathcal{L}(P)$, we employed two strategies for averaging. In the first case, values were averaged between all possible pairs of terms in $\mathcal{L}(T)$ and $\mathcal{L}(P)$. Specifically, we calculated $s(T, P)$ as

$$s(T, P) = \frac{1}{|\mathcal{L}(T)| \cdot |\mathcal{L}(P)|} \sum_{t \in \mathcal{L}(T)} \sum_{p \in \mathcal{L}(P)} s(t, p). \tag{2.19}$$

We refer to this form of averaging as *all-pair* averaging. This method of averaging was applied by Lord et al. [9] in calculating similarity between two functional annotations.

In the second case we calculated the similarity between the two sets as the average of the maximum similarity between a term from one set and all terms in the other. Specifically, we calculated $s(T, P)$ as

$$s(T, P) = \frac{1}{2|\mathcal{L}(T)|} \sum_{t \in \mathcal{L}(T)} \max_{p \in \mathcal{L}(P)} \{s(t, p)\} + \frac{1}{2|\mathcal{L}(P)|} \sum_{p \in \mathcal{L}(P)} \max_{t \in \mathcal{L}(T)} \{s(t, p)\}. \tag{2.20}$$

This measure represents the technique of averaging employed by Verspoor et al. [13] when calculating precision and recall (originally referred to as hierarchical precision and recall). There, the authors separately calculate precision as

$$pr(T, P) = \frac{1}{|\mathcal{L}(P)|} \sum_{p \in \mathcal{L}(P)} \max_{t \in \mathcal{L}(T)} \frac{|\mathcal{A}(t, p)|}{|\mathcal{A}(p)|} \tag{2.21}$$

and recall as

$$rc(T, P) = \frac{1}{|\mathcal{L}(T)|} \sum_{t \in \mathcal{L}(T)} \max_{p \in \mathcal{L}(P)} \frac{|\mathcal{A}(t, p)|}{|\mathcal{A}(t)|}. \tag{2.22}$$

We refer to this method of averaging as *max-average*. Although not implemented here, Schlicker et al. [12] employ a technique for averaging that is similar to Eq. (2.20), but takes the maximum average similarity for one set as opposed to the average between the two. Specifically,

2.1 Calculating the Joint Probability of a Graph

$$s(T, P) = \max \left\{ \frac{1}{|\mathcal{L}(T)|} \sum_{t \in \mathcal{L}(T)} \max_{p \in \mathcal{L}(P)} \{s(t, p)\}, \frac{1}{|\mathcal{L}(P)|} \sum_{p \in \mathcal{L}(P)} \max_{t \in \mathcal{L}(T)} \{s(t, p)\} \right\}. \tag{2.23}$$

When averaging pairwise comparisons for distance metrics, the above averages are calculated as the average of minimum pairwise distances instead of maximum pairwise similarities.

2.1.8 Additional Topological Metrics

In addition to information-theoretic metrics, we also used Jaccard's similarity coefficient [5] when calculating the similarity between the two consistent annotation graphs T and P. The Jaccard similarity coefficient is defined as

$$s(T, P) = \frac{|T \cap P|}{|T \cup P|}. \tag{2.24}$$

We note that cosine similarity as well as Maryland bridge coefficient [3] usually result in values correlated with the Jaccard similarity coefficient. For that reason, these two similarity measures were not presented.

2.2 Confusion Matrix Interpretation of ru and mi

While we introduce remaining uncertainty and misinformation in the context of comparing subgraphs of a larger ontology these two terms can also be intuitively interpreted as the information-theoretic analogs of false positives and false negatives or Type I and Type II errors. If we consider the confusion matrix in Table 2.2, we see the four positive outcomes that can occur when attempting to perform binary classification. False positives (Type I errors) are instances where a data point is a negative example but is incorrectly classified as a positive. In the context of hypothesis testing, these are cases where the Null hypothesis is true, but has incorrectly been rejected. False negatives (Type II errors) represent cases where the data point is a positive, but has incorrectly been labeled as a negative. These are cases where the Null hypothesis is false, but a model has failed to reject it.

Given a set of true terms, T, and predicted terms, P, then $P \setminus T$ would represent the terms that would fall in the false positive or Type I error portion of the confusion matrix. As defined in Eq. (2.6), this is the same set of terms whose joint probability, and subsequently information content, are measured when calculating misinformation. In this context, misinformation can be thought of as the number of bits of information, instead of traditional counts, the false positives represent. Conversely, if we consider false negatives, or the set of terms defined by $T \setminus P$ we are referring to

Table 2.2 A sample confusion matrix showing the four potential outcomes when performing binary classification

		True label	
		Positive	Negative
Predicted label	Positive	True positives	False positives (Type I)
	Negative	False negatives (Type II)	True negatives

the same set of terms used to calculate remaining uncertainty in Eq. (2.5). Remaining uncertainty, in this context, can be thought of as the number of bits of information, instead of traditional counts the false negatives represent.

2.3 Annotation Models

In order to judge the validity of our evaluation metrics, we implemented several well-studied annotation models. This was done by first collecting all proteins with GO annotations supported by experimental evidence codes (EXP, IDA, IPI, IMP, IGI, IEP, TAS, IC) from the January 2011 version of the Swiss-Prot database (29,699 proteins in MFO; 31,608 in BPO; and 30,486 in CCO). We then generated three simple function annotation models: Naïve, BLAST, and GOtcha, in order to assess the ability of performance metrics to accurately reflect the quality of a predicted set of annotations. In addition to these three methods, we generated another set of "predictions" by collecting experimental annotations for the same set of proteins from a database generated by the GO Consortium released at about the same time as our version of Swiss-Prot. This was done to quantify the variability of experimental annotation across different databases using the same set of metrics. In addition, this comparison can be used to estimate the empirical upper limit of prediction accuracy because the observed performance is limited by the noise in experimental data. All computational methods were evaluated using 10-fold cross-validation.

2.3.1 The Naïve Model

The Naïve model was designed to reflect biases in the distribution of terms in the data set and was the simplest annotation model we employed. It was generated by first calculating the relative frequency of each term in the training data set. This value was then used as the prediction score for every protein in the test set; thus, every protein in the test partition was assigned an identical set of predictions over all functional terms. The performance of the Naïve model reflects what one could expect when annotating a protein with no knowledge about that protein.

2.3.2 The BLAST Model

The BLAST model was generated using local sequence identity scores to annotate proteins. Given a target protein sequence x, a particular functional term v in the ontology, and a set of sequences $S_v = \{s_1, s_2, \ldots\}$ annotated with term v, we determine the BLAST predictor score for function v as

$$\max_{s \in S_v} \{sid(x, s)\},$$

where $sid(x, s)$ is the maximum sequence identity returned by the BLAST package [1] when the two sequences are aligned. We chose this method to mimic the performance one would expect if they simply used BLAST to transfer annotations between similar sequences.

2.3.3 The GOtcha Model

The third method, GOtcha [10], was selected to incorporate not only sequence identity between protein sequences, but also the structure of the ontology (technically, BLAST also incorporates structure of the ontology but in a relatively trivial manner). Specifically, given a target protein x, a particular functional term v, and a set of sequences $S_v = \{s_1, s_2, \ldots\}$ annotated with function v, one first determines the r-score for function v as

$$r_v = c - \sum_{s \in S_v} \log(e(x, s)),$$

where $e(x, s)$ represents the E-value of the alignment between the target sequence x and sequence s, and $c = 2$ is a constant added to the given quantity to ensure all scores were above 0. Given the r-score for function v, i-scores were then calculated by dividing the r-score of each function by the score for the root term $i_v = r_v/r_{\text{root}}$. As such, GOtcha is an inexpensive and robust predictor of function.

References

1. Altschul, S.F., et al.: Gapped BLAST and PSI-BLAST: a new generation of protein database search programs. Nucleic Acids Res. **25**(17), 3389–3402 (1997)
2. Clark, W.T., Radivojac, P.: Information-theoretic evaluation of predicted ontological annotations. Bioinformatics **29**(13), i53–i61 (2013)
3. Glazko, G., et al.: The choice of optimal distance measure in genome-wide datasets. Bioinformatics **21**(Suppl 3), iii3–iii11 (2005)
4. Guzzi, P.H., et al.: Semantic similarity analysis of protein data: assessment with biological features and issues. Briefings Bioinf. **13**(5), 569–585 (2012)

5. Jaccard, P.: Étude comparative de la distribution florale dans une portion des Alpes et des Jura. Bull. Soc. Vaud. des Sci. Nat. **37**, 574–579 (1901)
6. Jiang, J.J., Conrath, D.W.: Semantic similarity based on corpus statistics and lexical taxonomy. In: International Conference on Research in Computational Linguistics, pp. 19–33 (1997)
7. Koller, D., Friedman, N.: Probabilistic Graphical Models. The MIT Press, Cambridge (2009)
8. Lin, D.: An information-theoretic definition of similarity. In: Proceedings of the 15th International Conference on Machine Learning, pp. 296–304. Morgan Kaufmann, San Francisco (1998)
9. Lord, P.W., et al.: Investigating semantic similarity measures across the gene ontology: the relationship between sequence and annotation. Bioinformatics **19**(10), 1275–1283 (2003)
10. Martin, D.M., et al.: GOtcha: a new method for prediction of protein function assessed by the annotation of seven genomes. BMC Bioinf. **5**, 178 (2004)
11. Resnik, P.: Using information content to evaluate semantic similarity in a taxonomy. In: Proceedings of the 14th International Joint Conference on Artificial Intelligence, pp. 448–453 (1995)
12. Schlicker, A., et al.: A new measure for functional similarity of gene products based on gene ontology. BMC Bioinf. **7**, 302 (2006)
13. Verspoor, K., et al.: A categorization approach to automated ontological function annotation. Protein Sci. **15**(6), 1544–1549 (2006)

Chapter 3
Experiments and Results

In this chapter, we first analyze the average information content in a data set of experimentally annotated proteins and then evaluate performance accuracy of different function prediction methods using both topological and probabilistic metrics. Each experiment was conducted on all three categories of the Gene Ontology: Molecular Function (MFO), Biological Process (BPO), and Cellular Component (CCO) ontologies. In order to avoid cases where the information content of a term is infinite a pseudocount of 1 was added to each term, and the total number of proteins in the data set was incremented when calculating term frequencies.

3.1 Average Information Content of a Protein

We first examined the distribution of the information content per protein for each of the three ontologies (Fig. 3.1). We observe a wide range of information contents in all ontologies, reaching over 128 bits in case of BPO (which corresponds to a factor of 128 in the probability of observing particular annotation graphs). The distributions for MFO and CCO show unusual peaks for very low information contents, suggesting that a large fraction of annotation graphs in these ontologies are low quality. One such anomaly is created by the term "binding" in MFO that is associated with 72 % of proteins. Furthermore, 41 % of proteins are annotated with its child "protein binding" as a leaf term, and 26 % are annotated with it as their sole leaf term. Such annotations, which are clearly a consequence of high-throughput experiments, present a significant difficulty in method evaluation.

Previously, we showed that the distribution of leaf terms in protein annotation graphs exhibits scale-free tendencies [1]. Here, we also analyzed the average number of leaf terms per protein and compared it with the information content of that protein. We estimate the average number of leaf terms to be 1.6 (std. 1.0), 3.0 (std. 3.6), and 1.6 (std. 1.0) for MFO, BPO, and CCO, respectively, and calculate Pearson correlation between the information content and the number of leaf terms for a protein (0.80,

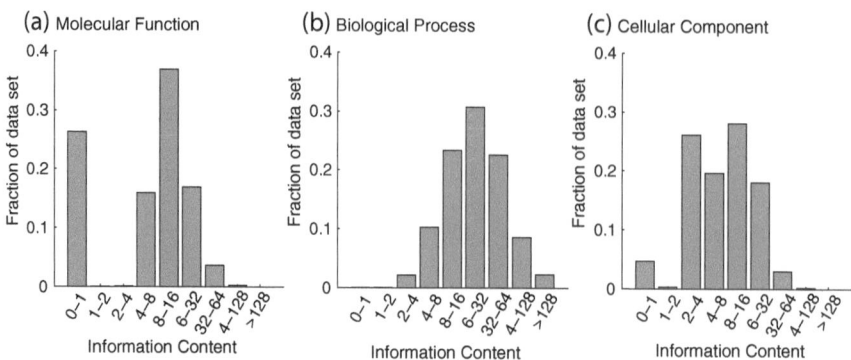

Fig. 3.1 Distribution of information content (in bits) over proteins annotated by terms for each of the three ontologies. The average information content of a protein was estimated at 10.9 (std. 10.2), 32.0 (std. 33.6), and 10.4 (std. 9.2) bits for MFO, BPO, and CCO, respectively

0.92, and 0.71). Such high level of correlation suggests that proteins annotated with a small number of leaf terms are generally annotated by shallow graphs. This is particularly evident in the case of "protein binding" annotations that can be derived from yeast-2-hybrid experiments, but provide little insight into the functional aspects of these complexes when only viewed as GO annotations. We believe the wide range of information contents coupled with the fact that a large fraction of proteins were essentially uninformative, justifies the weighting proposed in this work.

3.2 Comparative Examples of Calculating Information Content

One of the major strengths of our metrics is they are founded in sound methodology for calculating the information content of a set of terms. In order to illustrate the importance of this, we selected two proteins from our data set: SOT18_ARATH and M3K1_RAT.

Each protein is annotated with several leaf annotations shown in Table 3.1. In the version of Swiss-Prot used, SOT18_ARATH is annotated with several direct children of desulfoglucosinolate sulfotransferase activity. M3K1_RAT is annotated with several terms relating to kinase activity and binding, ligase activity, and lipid binding. As can be seen by looking at each protein's annotation graphs, although there are terms that are closely related, M3K1_RAT's annotations define a much wider annotation graph than SOT18_ARATH's (Fig. 3.2).

We first calculate the information content of each protein's annotation by treating each leaf term as independent and summed their information content. In doing so we obtain values that are very close: 83.18 bits and 82.94 bits for protein SOT18_ARATH and M3K1_RAT, respectively. One would expect that an annotation graph that defines a wider range of terms would be more informative than one that is defined by the

3.2 Comparative Examples of Calculating Information Content

Table 3.1 Experimental leaf annotations of SOT18_ARATH and M3K1_RAT taken from the January 2011 version of Swiss-Prot

SOT18_ARATH	M3K1_RAT
$i = 83.18$ bits	$i = 82.94$ bits
$i_{bn} = 18.86$ bits	$i_{bn} = 62.19$ bits
indol-3-yl-methyl-desulfoglucosinolate sulfotransferase activity	sphingolipid binding
8-methylthiooctyl-desulfoglucosinolate sulfotransferase activity	mitogen-activated protein kinase kinase binding
7-methylthioheptyl-desulfoglucosinolate sulfotransferase activity	JUN kinase binding
5-methylthiopentyl-desulfoglucosinolate sulfotransferase activity	mitogen-activated protein kinase binding
4-methylthiobutyl-desulfoglucosinolate sulfotransferase activity	identical protein binding
3-methylthiopropyl-desulfoglucosinolate sulfotransferase activity	ubiquitin-protein ligase activity
	MAPERK kinase kinase activity
	JUN kinase kinase kinase activity

The information content of these annotations was calculated by summing the information content of each leaf term (i), and by using the structure of the ontology to calculate the joint probability of all terms in the annotation sub graph (i_{bn})

same number of closely related terms. This is clearly reflected when we use Eq. (2.4b) to calculate the joint probability, and subsequently information content of all terms. The information content of each protein's set of annotations we come up with much more intuitive values, 18.86 bits and 62.19 bits for SOT18_ARATH and M3K1_RAT respectively. In both cases there is an observed decrease in information content, but this decrease is more notable for SOT18_ARATH.

Our method for calculating information content follows the intuition that annotations with terms that occur far apart in the ontology contain more information, whereas annotations consisting of terms that appear very close to each other are less informative. Metrics that rely on different methods for averaging pairwise comparisons between terms are problematic because they implicitly assume the terms that comprise an annotation are independent. As illustrated by SOT18_ARATH's annotations it is often the case that annotations can consist of multiple variations of similar functions.

One could also envision calculating the information content of a set of annotations independent of the ontology. This thought exercise illustrates the utility of using the predefined structure of the ontology to designate relationships between terms. In such a situation information content could be calculated by considering the frequency with which that exact set of terms is assigned to a protein from a data set. It is highly likely, especially in cases where annotations are remotely detailed, that the specific given combination of terms is uniquely assigned to only a single protein; resulting in an information content value of $-\log(N-1)$, where N represents the number of proteins

32   3 Experiments and Results

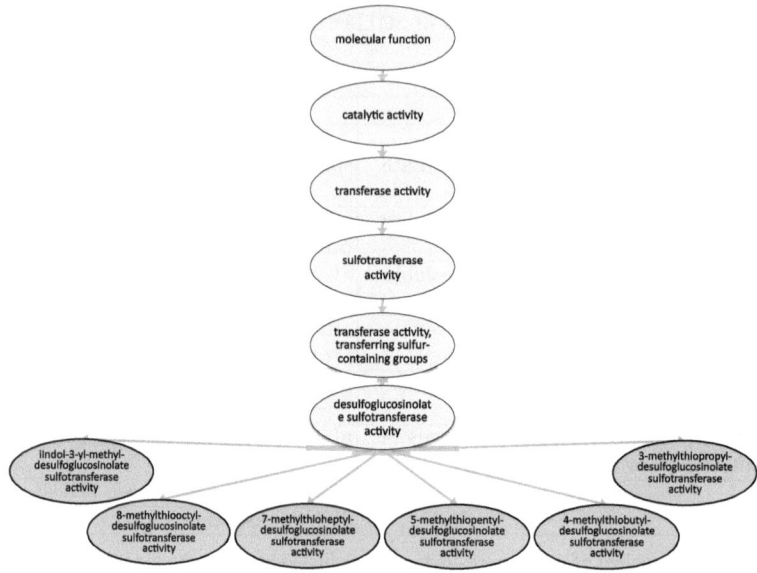

(a) SOT18_ARATH

(b) M3K1_RAT

Fig. 3.2 The annotation graphs for SOT18_ARATH and M3K1_RAT. *Dashed edges* indicate intermediate terms have been removed in order to save space

3.2 Comparative Examples of Calculating Information Content

in the data set. Under the ontology-free model a large fraction of annotations could have high information content associated with them, regardless of how detailed the annotations are, if only a few data points have the exact same annotation.

3.3 Two-Dimensional Plots

In order to assess how each metric evaluated the performance of the four prediction methods, we generated two-dimensional plots. Figure 3.3 shows the performance of each predictor using precision/recall and ru–mi curves, as well as their weighted variants. The performance of the GO/Swiss-Prot annotation is represented as a single point because it compares two databases of experimental annotations where predictions are all binary and do not have associated scores.

When looking at the precision–recall curves, we first observe an unusually high area under the curve associated with the Naive model. This is a result of a significant fraction of low information content annotations that are relatively easy to predict by simply using prior probabilities of terms as prediction values. In addition, these biases lead to a biologically unexpected result where the predictor based on the BLAST algorithm performs on par with the Naive model, e.g., $F_{\max}$ (BLAST, MFO) = 0.65 and $F_{\max}$ (Naive, MFO) = 0.60, while $F_{\max}$ (BLAST, CCO) = 0.63; $F_{\max}$ (Naive, CCO) = 0.64. The largest difference between the BLAST and Naive models was observed for BPO, which has a Gaussian-like distribution of information contents in the logarithmic scale (Fig. 3.1). The second column of plots in Fig. 3.3 shows the weighted precision–recall curves. Here, we observe large changes in the performance accuracy, especially for the Naive model, in MFO and CCO categories, whereas the BPO category was, for the most part, not impacted. We believe that the information-theoretic weighting of precision and recall resulted in more meaningful evaluation.

In Fig. 3.4, we present more detailed results related to the ru–mi curves presented in Fig. 3.3. In the top row, we show the same ru–mi curves as in the third column of Fig. 3.3, with yellow dots providing information where the maximum values of semantic distance S_2 were reached for each method. Interestingly, because the predictors generally associate scores to all nodes in the Gene Ontology, the amount of overprediction can be very large for low decision thresholds, which consequently results in large values of misinformation. To provide better insight into the balance between remaining uncertainty and misinformation achieved by the semantic distance S_2, in the bottom row we present the same curves for small values of misinformation only.

The information-theoretic measures are shown in the last two columns of Fig. 3.3 and in Fig. 3.4. One useful property of ru–mi plots is that they explicitly illustrate how many bits of information are yet to be revealed about a protein (on average) as a function of misinformation that is introduced by overprediction or misannotation. In all three categories, the amount of misinformation being introduced increases rapidly; quickly obtaining a rate that is twice the amount of expected information for an average protein. We believe these plots shed new light into how much information

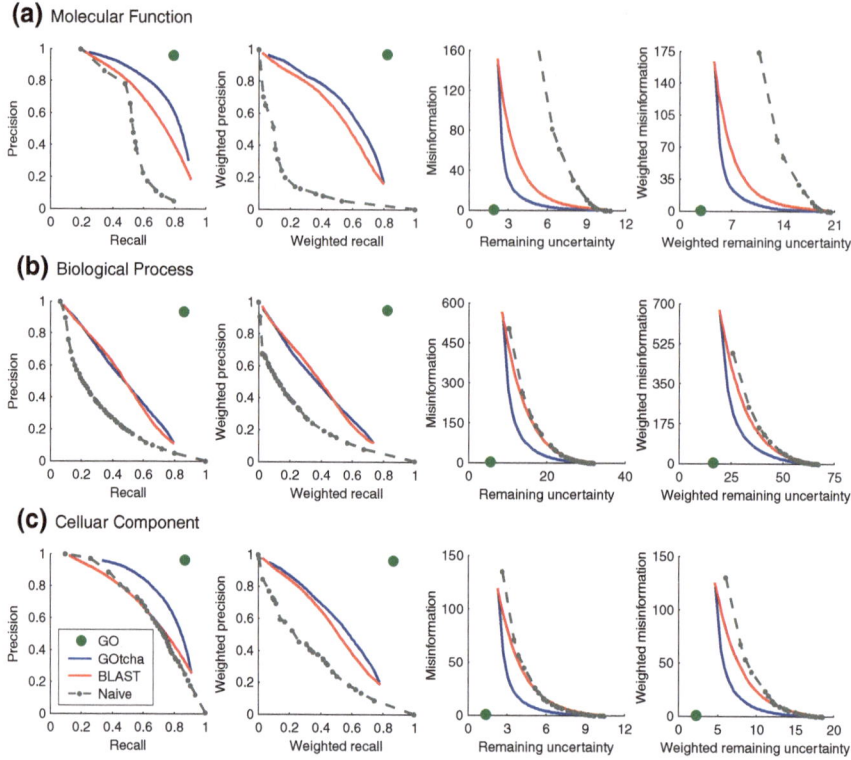

Fig. 3.3 Two-dimensional evaluation plots. Each plot shows three prediction methods: Naive (*grey, dashed*), BLAST (*red, solid*), and GOtcha (*blue, solid*) constructed using cross-validation. *Green point*-labeled GO shows the performance evaluation between two databases of experimental annotations, downloaded at the same time. The *rows* show the performance for different ontologies (MFO, BPO, CCO). The *columns* show different evaluation metrics: $(pr(\tau), rc(\tau))_\tau$, $(wpr(\tau), wrc(\tau))_\tau$, $(ru(\tau), mi(\tau))_\tau$, and $(wru(\tau), wmi(\tau))_\tau$

overload a researcher can be presented with by drawing predictions at a particular threshold. Looking from right to left in each plot, we observe an elbow in each of the curves (at about 3 bits for MFO and CCO and 12 bits for BPO; Fig. 3.3) after which the remaining uncertainty barely decreases, while misinformation grows out of control.

Figures 3.5 and 3.6 show results when implementing the supplemental information-theoretic metrics using all-pair and max-average methods of averaging, respectively. It is important to mention that in a direct application of Resnik's similarity function, we refer to it as *Lord* in Fig. 3.5 (all-pair averaging) and as *Resnik* in Fig. 3.6 (max-average method of averaging). This is because, to the best of our knowledge, in the context of comparing functional annotations of proteins the all-pair averaging was first proposed by Lord et al. [4].

3.3 Two-Dimensional Plots

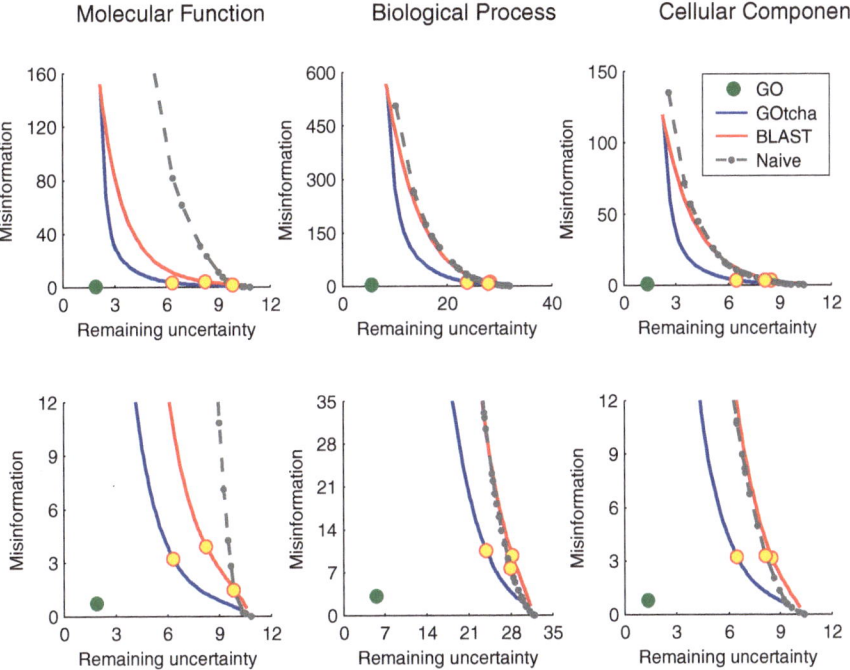

Fig. 3.4 Figures showing the remaining uncertainty and misinformation of baseline methods with *yellow dots* denoting values at which each method obtains its S_2 value. For better interpretation of the values at which each method achieves its S_2 value the *bottom row* of figures show the same plots as the *top row*, but with adjusted y-axis limits

In Fig. 3.7 we contrast two different types of precision–recall curves. In the top row, we present the same results as in Fig. 3.1 of the main manuscript. In the bottom row, we use the max-averaging method for calculating precision–recall outlined by Verspoor et al. [7]. The methods provided similar results although the max-average formulation had generally larger values of F_{max} than the standard formulation of precision and recall (as defined in the main manuscript). However, these larger values of F_{max} occurred at lower decision thresholds.

3.4 Comparisons of Single Statistics

Here we analyze the ability of single measures to rank predictors and lead to useful evaluation insights. We compare the performance of semantic distance to several other methods that calculate either topological or semantic similarities. For each evaluation method, the decision threshold was varied for each of the prediction methods

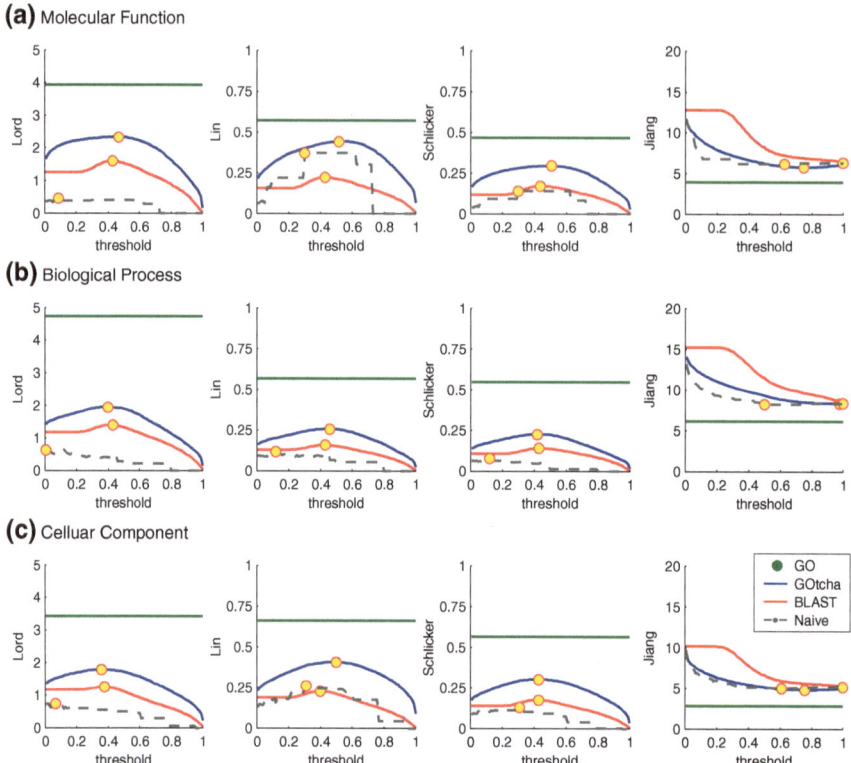

Fig. 3.5 Two-dimensional evaluation plots of information content-based metric performances when using the all-pair method of averaging. *Yellow dots* denote the maximum similarity or, in the case of [2], the minimum distance, that each method obtains

and the threshold providing the best performance was selected as optimal. We then analyze and discuss the performance of these metrics at those optimal thresholds.

We implemented the semantic similarity metrics of [2, 3, 5, 6], as detailed in Sect. 2.1.7. Because each of these measures is only defined for a single pair of terms in the ontology, scores between two protein annotation graphs (true graph T vs. predicted graph P) we implemented both all pair averaging (Tables 3.2 and 3.3) and max-averaging (Table 3.4) for these metrics as detailed in Sect. 2.1.7.3. We note that the all-pair averaging using Resnik's term similarity was implemented by Lord et al. [4] in the context of GO annotations. In addition to these semantic measures, we also implemented the Jaccard similarity coefficient between the sets of vertices in the two annotation graphs (Sect. 2.1.8). For precision–recall curves and ru–mi curves, we used F_{max} and S_2 measures as single values to obtain optimal thresholds.

Tables 3.2, 3.3, and 3.4 show the maximum similarity, or minimum distance in the case of Jiang and Conrath's and semantic distance, that each metric obtained for each of our classification models. In addition to reporting the maximum similarity,

3.4 Comparisons of Single Statistics

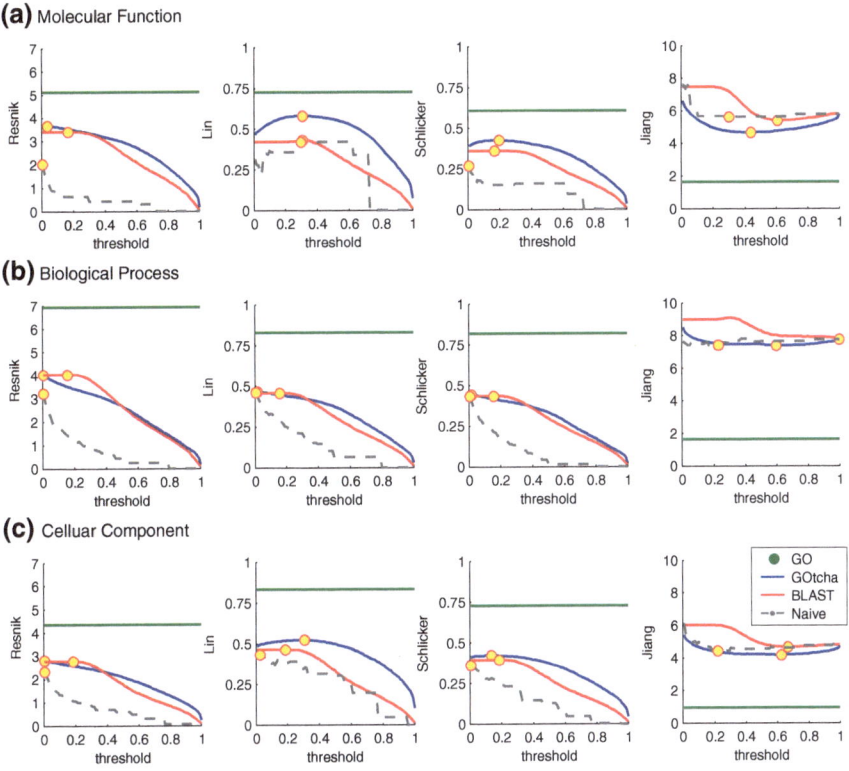

Fig. 3.6 Two-dimensional evaluation plots of information content-based metric performances when using the max-average method of averaging. *Yellow dots* denote the maximum similarity or, in the case of [2], the minimum distance, that each method obtains

we also report the decision threshold at which that value was obtained along with the associated level of remaining uncertainty and misinformation at that threshold.

Considering Tables 3.2 and 3.3 the first interesting observation is that all metrics, aside from that of Jiang and Conrath, obtain optimal thresholds that result in relatively similar levels of remaining uncertainty and misinformation for the GOtcha model. However, all metrics, aside from semantic distance and Jiang and Conrath's distance seem to favor extremely high levels of misinformation at the reported decision thresholds for the BLAST model. For MFO and CCO, the semantic similarity measures of Lord et al., Lin and Sclicker et al. report misinformation levels that are more than twice the information content of the average protein in that ontology for the BLAST model. In BPO those are even more extreme. We believe this is a direct consequence of the pairwise term averaging applied in these methods.

In Table 3.4, we provide results analogous to those from Table 3.2, but using max-average method instead of all-pair averaging. We generally observe similar trends as before, but note that the values of BLAST thresholds used for functional transfer are

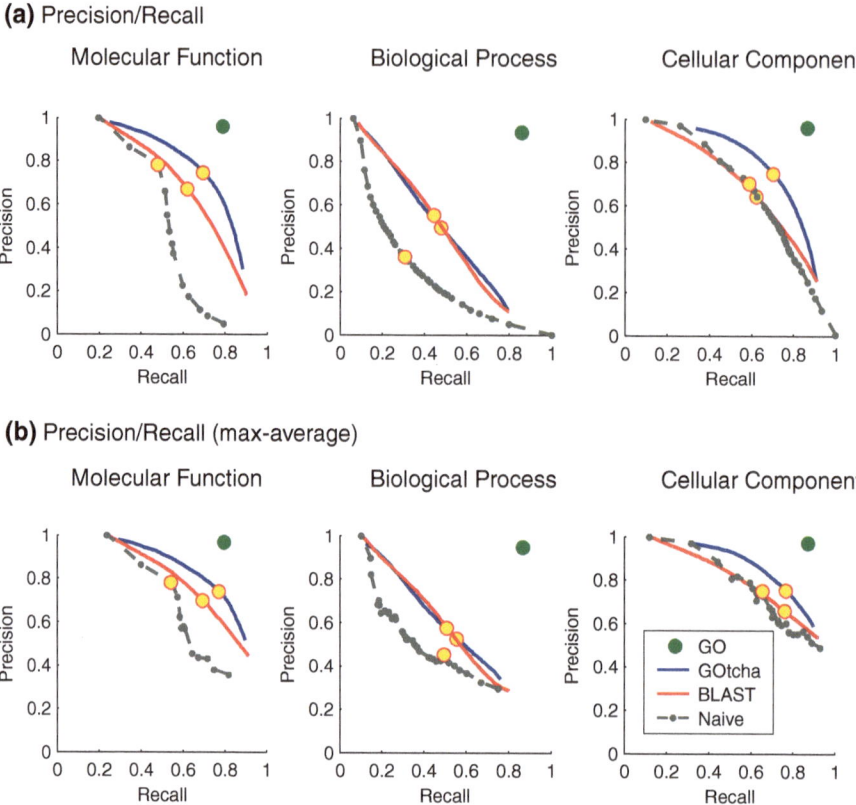

Fig. 3.7 Plots showing results using the standard method of calculating precision and recall (*top row*) compared to using the max-average method of calculating precision and recall as detailed by Verspoor et al. [7] (*bottom row*). *Yellow dots* denote the values of precision and recall at which each method obtains its F_{max} value

even lower than when all-pair averaging was used (except for the measure by Jiang and Conrath [2], where max-average method seems to be beneficial).

It is particularly interesting to analyze the optimal thresholds obtained for the BLAST model. These thresholds can be interpreted as the level of sequence identity above which each metric reports functional transfer can be made. For example, because their optimal BLAST thresholds are relatively low, the levels of misinformation provided by the similarities of Lord et al., Lin, and Schlicker et al. are rather large in both tables. F_{max} and Jaccard approaches also report low threshold values for all ontologies, while Jiang and Conrath's distance selects the optimal threshold at an overly restrictive 100% sequence identity. We believe that the semantic distance S_2 provides more reasonable values for functional transfer, finding an optimal distance at 77, 88, and 78% for MFO, BPO, and CCO, respectively.

3.4 Comparisons of Single Statistics

Table 3.2 Performance evaluation of common information-theoretic metrics

Lord	Molecular Function				Biological Process				Cellular Component			
	Max	Threshold	ru	mi	Max	Threshold	ru	mi	Max	Threshold	ru	mi
GOtcha	2.34	0.47	6.34	3.20	1.95	0.40	23.36	11.90	1.80	0.36	5.88	4.58
BLAST	1.61	0.43	4.69	27.90	1.40	0.43	16.73	139.57	1.27	0.38	4.42	37.24
Naive	0.46	0.09	9.56	4.23	0.63	0.01	10.35	504.88	0.75	0.07	5.81	16.34

Lin	Max	Threshold	ru	mi	Max	Threshold	ru	mi	Max	Threshold	ru	mi
GOtcha	0.44	0.52	6.67	2.67	0.26	0.46	24.43	9.40	0.41	0.50	6.71	2.76
BLAST	0.22	0.43	4.69	27.90	0.16	0.43	16.73	139.57	0.23	0.40	4.78	30.45
Naive	0.37	0.30	10.39	0.21	0.12	0.12	24.92	23.14	0.26	0.31	8.98	1.32

Schlicker	Max	Threshold	ru	mi	Max	Threshold	ru	mi	Max	Threshold	ru	mi
GOtcha	0.29	0.51	6.60	2.76	0.23	0.42	23.73	10.99	0.30	0.43	6.31	3.56
BLAST	0.17	0.44	4.83	25.39	0.14	0.43	16.73	139.57	0.18	0.43	5.26	23.26
Naive	0.14	0.30	10.39	0.21	0.08	0.12	24.92	23.14	0.13	0.31	8.98	1.32

Jiang	Min	Threshold	ru	mi	Min	Threshold	ru	mi	Min	Threshold	ru	mi
GOtcha	5.74	0.75	8.20	1.27	8.38	0.98	30.88	1.22	4.83	0.76	8.21	1.19
BLAST	6.34	1.00	10.62	0.43	8.39	1.00	31.31	1.40	5.20	1.00	10.16	0.35
Naive	6.19	0.63	10.53	0.13	8.24	0.50	31.75	0.07	5.01	0.61	10.13	0.08

For each measure, the decision threshold was varied across the entire range of predictions to obtain the maximum or minimum value (shown in column 1). The threshold at which each method reached the best value is shown in column 2. Columns 3 and 4 show the remaining uncertainty (ru) and misinformation (mi) calculated according to the Bayesian network. Each semantic similarity metric was calculated according to the relative frequencies of observing each term in the database

Table 3.3 Performance evaluation of topological metrics and S_2

Jaccard	Molecular Function				Biological Process				Cellular Component			
	Max	Threshold	ru	mi	Max	Threshold	ru	mi	Max	Threshold	ru	mi
GOtcha	0.57	0.46	6.29	3.32	0.31	0.34	22.24	15.24	0.56	0.43	6.31	3.56
BLAST	0.37	0.50	5.74	14.72	0.19	0.50	19.68	76.98	0.34	0.43	5.26	23.26
Naive	0.46	0.30	10.39	0.21	0.17	0.19	27.53	9.22	0.47	0.31	8.98	1.32

F_{max}	Max	Threshold	ru	mi	Max	Threshold	ru	mi	Max	Threshold	ru	mi
GOtcha	0.72	0.43	6.12	3.68	0.49	0.32	21.84	16.69	0.73	0.43	6.31	3.56
BLAST	0.64	0.48	5.42	17.89	0.49	0.50	19.68	76.98	0.63	0.45	5.57	19.42
Naive	0.60	0.29	9.87	1.44	0.33	0.19	27.53	9.22	0.64	0.33	9.22	0.80

S_2	Min	Threshold	ru	mi	Min	Threshold	ru	mi	Min	Threshold	ru	mi
GOtcha	7.11	0.47	6.34	3.20	26.14	0.43	23.91	10.56	7.23	0.46	6.48	3.21
BLAST	9.13	0.77	8.25	3.90	29.89	0.88	28.28	9.69	9.08	0.78	8.51	3.15
Naive	9.98	0.10	9.72	2.80	29.00	0.22	27.67	8.72	8.79	0.21	7.71	4.95

For each measure, the decision threshold was varied across the entire range of predictions to obtain the maximum or minimum value (shown in column 1). The threshold at which each method reached the best value is shown in column 2. Columns 3 and 4 show the remaining uncertainty (ru) and misinformation (mi) calculated according to the Bayesian network. Each semantic similarity metric was calculated according to the relative frequencies of observing each term in the database

Table 3.4 Performance of information-theoretic methods when calculating performance as the average of maximum similarity (or distance) between each true and predicted term as described in Sect. 2.8.3

	Molecular Function				Biological Process				Cellular Component			
Resnik	Max	Threshold	ru	mi	Max	Threshold	ru	mi	Max	Threshold	ru	mi
GOtcha	3.65	0.04	2.96	32.17	4.02	0.01	10.19	273.90	2.80	0.01	2.70	59.02
BLAST	3.41	0.17	2.17	151.77	4.03	0.16	8.48	566.88	2.77	0.19	2.26	119.34
Naive	2.02	0.01	5.07	177.91	3.22	0.01	10.35	504.88	2.32	0.01	2.65	135.04
Lin	Max	Threshold	ru	mi	Max	Threshold	ru	mi	Max	Threshold	ru	mi
GOtcha	0.58	0.31	5.32	5.86	0.46	0.02	11.10	192.89	0.52	0.31	5.55	5.58
BLAST	0.43	0.31	2.90	90.72	0.45	0.16	8.48	566.88	0.46	0.19	2.26	119.34
Naive	0.42	0.30	10.39	0.21	0.46	0.01	10.35	504.88	0.43	0.03	3.90	56.83
Jiang	Min	Threshold	ru	mi	Min	Threshold	ru	mi	Min	Threshold	ru	mi
GOtcha	4.65	0.44	6.18	3.56	7.39	0.60	26.65	5.30	4.19	0.63	7.45	1.80
BLAST	5.38	0.61	6.97	7.58	7.75	1.00	31.31	1.40	4.67	0.67	7.84	5.13
Naive	5.61	0.30	10.39	0.21	7.41	0.23	27.73	8.49	4.45	0.22	8.17	3.25
Schlicker	Max	Threshold	ru	mi	Max	Threshold	ru	mi	Max	Threshold	ru	mi
GOtcha	0.42	0.20	4.56	9.32	0.44	0.02	11.10	192.89	0.42	0.14	4.29	12.76
BLAST	0.36	0.17	2.17	151.77	0.43	0.16	8.48	566.88	0.39	0.19	2.26	119.34
Naive	0.27	0.01	5.07	177.91	0.43	0.01	10.35	504.88	0.36	0.01	2.65	135.04
F_{max}	Max	Threshold	ru	mi	Max	Threshold	ru	mi	Max	Threshold	ru	mi
GOtcha	0.76	0.33	5.45	5.41	0.54	0.23	19.79	26.04	0.76	0.31	5.55	5.58
BLAST	0.70	0.44	4.83	25.39	0.54	0.46	18.06	108.73	0.71	0.37	4.23	41.17
Naive	0.64	0.29	9.87	1.44	0.47	0.07	22.06	51.55	0.70	0.20	7.23	6.75

F_{max} values were calculated using precision and recall as detailed by Verspoor et al. [7]. The decision threshold was varied across the entire range of predictions to obtain the maximum or minimum value (shown in column 1) for each method. The threshold at which each method reached the best value is shown in column 2. Columns 3 and 4 show the remaining uncertainty and misinformation calculated according to the Bayesian network

We believe that, generally, all-pair averaging provides better results regarding functional similarity than does max-average. This is predominantly based on the even lower optimal thresholds obtained for the BLAST model in Table 3.4.

References

1. Clark, W.T., Radivojac, P.: Analysis of protein function and its prediction from amino acid sequence. Proteins: Struct., Funct., Bioinf. **79**(7), 2086–2096 (2011)
2. Jiang, J.J., Conrath, D.W.: Semantic similarity based on corpus statistics and lexical taxonomy. In international conference on research in computational linguistics, 19–33, 1997
3. Lin, D.: An information-theoretic definition of similarity. In: Proceedings of the 15th International Conference on Machine Learning, pp. 296–304. Morgan Kaufmann (1998)
4. Lord, P.W., et al.: Investigating semantic similarity measures across the Gene Ontology: the relationship between sequence and annotation. Bioinformatics **19**(10), 1275–1283 (2003)

References

5. Resnik, P.: Using information content to evaluate semantic similarity in a taxonomy. In: Proceedings of the 14th International Joint Conference on Artificial Intelligence, 448–453, 1995
6. Schlicker, A., et al.: A new measure for functional similarity of gene products based on Gene Ontology. BMC Bioinformatics **7**, 302 (2006)
7. Verspoor, K., et al.: A categorization approach to automated ontological function annotation. Protein Sci. **15**(6), 1544–1549 (2006)

Chapter 4
Discussion

Here, we introduce an information-theoretic framework for evaluating the performance of computational protein function prediction. We frame protein function prediction as a structured output learning problem in which the output space is represented by consistent subgraphs of the GO graph. We argue that our approach directly addresses evaluation in cases where there are multiple true and predicted (leaf) terms associated with a protein by taking the structure of the ontology and the dependencies between terms induced by a hierarchical ontology into account. Our method also facilitates accounting for the high level of biased and incomplete experimental annotations of proteins by allowing for the weighting of proteins based on the information content of their annotations. Because we maintain an information-theoretic foundation, our approach is relatively immune to the potential dissociation between the depth of a term and its information content; a weakness of often-used topological metrics in this domain such as precision/recall or ROC-based evaluation. At the same time, because we take a holistic approach to considering a protein's potentially large set of true or predicted functional associations, we resolve many of the problems introduced by the practice of aggregating multiple pairwise similarity comparisons common to existing semantic similarity measures.

Although there is a long history [3] and a significant body of work in the literature regarding the use of semantic similarity measures [1, 2], to the best of our knowledge all such metrics are based on single statistics and are unable to provide insight into the levels of remaining uncertainty and misinformation that every predictor is expected to balance. Therefore, the methods proposed in this work extend, modify, and formalize several useful information-theoretic metrics introduced over the past decades. In addition, both remaining uncertainty and misinformation have natural information-theoretic interpretations and can provide meaningful information to the users of computational tools. At the same time, the semantic distance based on these concepts facilitates not only the use of a single performance measure to evaluate and rank predictors, but can also be exploited as a loss function during training.

One limitation of the proposed approach is grounded in the assumption that a Bayesian network, structured according to the underlying ontology, will perfectly model the prior probability distribution of a target variable. An interesting anomaly with this approach is that the marginal probability, and subsequently the information content, of a single term (i.e., consistent graph with a single leaf term) calculated by considering the conditional probabilities between child and parent terms in the graph does not necessarily match the relative term frequency in the database. Ad hoc solutions that maintain the term information content are possible but would result in sacrificed interpretability of the metric itself. One such solution can be obtained via a recursive definition for Information accretion

$$ia(v) = i(v) - \sum_{u \in \mathcal{P}(v)} ia(u)$$

where $i(v)$ is estimated directly from the database and $ia(root) = 0$.

Finally, rationalizing between evaluation metrics is a difficult task. The literature presents several strategies where protein sequence similarity, protein–protein interactions or other data are used to assess whether a performance metric behaves according to expectations [1]. In this work, we took a somewhat different approach and showed that the demonstrably biased protein function data can be shown to provide surprising results with well-understood prediction algorithms and conventional evaluation metrics. Thus, we believe that our experiments provide evidence of the usefulness of the new evaluation metric.

References

1. Guzzi, P.H., et al.: Semantic similarity analysis of protein data: assessment with biological features and issues. Briefings Bioinf. **13**(5), 569–585 (2012)
2. Pesquita, C., et al.: Semantic similarity in biomedical ontologies. PLoS Comput. Biol. **5**(7), e1000443 (2009)
3. Resnik, P.: Semantic similarity in a taxonomy: an information-based measure and its application to problems of ambiguity in natural language. J. Artif. Intell. Res. **11**, 95–130 (1999)

Index

A
Annotation, 6
   models, 26
   predicted, 17
   transfer of, 7, 27
   true, 17
Averaging, 31
   all pair, 24
   maximum, 24

B
Bayesian network, 13
BLAST, 27, 37, 38
   E-value, 27

C
Common ancestor
   most informative, 18, 23
Conditional probability, 14
   distribution, 15
Confusion matrix, 25
Consistency, 13
Cross-validation, 26

D
Decision threshold, 18

E
Euclidean distance, 20

F
F-measure, 21, 33

False negative, 25
False positive, 25

G
Gene ontology, 1, 26
Gene prioritization, 6
Graphs
   annotation, 13

H
Harmonic mean, 21
Homology, 7
Hypothesis testing, 25
   Type I error, 25
   Type II error, 25

I
Information accretion, 16, 44
Information content, 16
   average, 29
   of an individual term, 23

J
Jaccard similarity, 25
Joint probability, 14

L
Leaf term, 4, 23, 29

M
Manhattan distance, 20

Markov blanket, 15
Microarray, 7
Misinformation, 18, 33
   average, 19
   weighted, 20

N
Null hypothesis, 25

O
Ontology, 1
Orthology, 7

P
Phylogenetics, 7
Physicochemical properties, 7
Precision, 21, 33
   weighted, 21
Protein function, 3

Protein structures, 7
Protein–protein interaction networks, 7
PSI-BLAST, 7

R
Recall, 21, 33
   weighted, 21
Remaining uncertainty, 17, 33
   average, 19
   weighted, 20

S
Semantic distance, 20, 33
Semantic similarity, 18, 23–24, 34
Sequence alignment, 7
Sequence identity, 7
   local, 27
Supervised learning, 7
Swiss-Prot, 26